FAUNA ENTOMOLOGICA SCANDINAVICA
Volume 20   1987

The aquatic Adephaga (Coleoptera) of Fennoscandia and Denmark

I.
Gyrinidae, Haliplidae, Hygrobiidae and Noteridae

by

Mogens Holmen

E. J. Brill/Scandinavian Science Press Ltd.

Leiden · Copenhagen

© *Copyright*
Scandinavian Science Press 1987

Fauna entomologica scandinavica
is edited by "Societas entomologica scandinavica"

Editorial board
Nils M. Andersen, Karl-Johan Hedqvist, Hans Kauri,
N. P. Kristensen, Harry Krogerus, Leif Lyneborg,
Hans Silfverberg

Managing editor
Leif Lyneborg

World list abbreviation
Fauna ent. scand.

Colour separation
Grafodan, Værløse, Denmark

Printed by
Vinderup Bogtrykkeri A/S
7830 Vinderup, Denmark

ISBN 9004.08185.2
ISBN 87-87491-39-7
ISSN 0106-8377

Author's address:
Zoological Museum
Universitetsparken 15
DK-2100 Copenhagen
Denmark

Contents

Introduction

The present volume treats four small families of aquatic beetles belonging to the so-called Coleoptera Adephaga, viz., Gyrinidae (whirligig beetles), Haliplidae (crawling water-beetles), Hygrobiidae (squeaker beetles) and Noteridae (burrowing water-beetles).

The Danish and Fennoscandian faunas of these families comprise 13 species of Gyrinidae, 21 species of Haliplidae and 2 species of Noteridae, making a total of 36 species. All North European species of the four families are included, adding a further 3 species of Gyrinidae, 1 species of Hygrobiidae and 1 species of Haliplidae.

The volume starts with keys to the Fennoscandian families of Adephaga for both adults and larvae. General information on morphology, zoogeography and bionomics is given under each family.

The keys and descriptions have generally been kept rather long, allowing the use of as many diagnostic characters as possible, many species being difficult to separate without examination of male and/or female genital sclerites.

The keys and the descriptions of species are only sufficient for the separation of North European species, but the remaining sections treat the families on a world basis.

Some revisional work has been included. The following two new synonymies have been established: *Haliplus apicalis* Thomson = *H. strigatus* Roberts, and *H. lineolatus* Mannerheim = *H. schaumi* Solsky. Lectotypes have been designated for the following species: *Dytiscus impressus* Fabricius, *Haliplus affinis* Stephens, *H. sibiricus* Motschulsky, *H. schaumi* Solsky, *H. nomax* Balfour-Browne, *H. browneanus* Sharp, and *H. interjectus* Lindberg.

Acknowledgements

I wish to express my gratitude towards the large number of persons who have willingly placed specimens and information at my disposal or in other ways have contributed to this volume. Special thanks are due to G. Andersson, Göteborg; B. Andrén, Svanesund; R. Baranowski, Lund; O. Biström, Helsinki; R. Danielsson, Lund; G. N. Foster, Prestwick; P. Gjelstrup, Århus; N. Haarløv, Copenhagen; G. Hagenlund, Oslo; M. Hansen, Copenhagen; L. H. Herman, New York; F. Hieke, Berlin (DDR); H. Hippa, Helsinki; L. Huldén, Helsinki; C. F. Jensen, Århus; L. Greve Jensen, Bergen; P. C. Jeppesen, Copenhagen; S. Jonsson, Uppsala; A. Fjeldså, Bergen; M. D. Kerley, London; A. G. Kirejtshuk, Leningrad; P. Lindskog, Stockholm; E. T. Linnaluoto, Turku; A. N. Nilsson, Umeå; B. Overgaard, Århus; T. Palm, Uppsala; S. Persson, Landskrona; J. E. Raastad, Oslo; G. E. Rotheray, Edinburgh; G. Scherer, München; A. G. Shatrovskyi, Leningrad; H. Silfverberg, Helsinki; J. van Tol, Leiden; B. van Vondel, Hendrik Ido Ambacht; J. Økland, Oslo.

I am very grateful to L. Lyneborg, Copenhagen and O. Lomholdt, Copenhagen for very valuable critisism of my work, to M. Luff, Newcastle, for the linguistic correction of the manuscript, to S. Brantlová, Prague, for preparing the colour plate, and to B. W. Rasmussen, Copenhagen, for preparing the electroscan micrographs. Also I am much indepted to my family and friends for encouragement and patience during this work.

Finally, I wish to thank the boards of the following foundations for financial aid to this study: "A. D. Clements Legat", "Dr. Meinerts and Hustrues Legat" and "Dr. phil. Axel M. Hemmingsens Legat".

The families of suborder Adephaga

The families of Adephaga may be recognized by the following combination of characters (Crowson, 1955). Adult: hind coxae immovably fused to metasternum, completely dividing first visible abdominal sternite; 6 visible abdominal sternites (those of segments 2-7) usually present, the first 3 sternites being connate; prothorax nearly always with notopleural suture distinct; hind wings usually with an "Oblongum Cell"; and aedeagus with parameres never attached to a basal piece. Larva: legs normally with a distinct tarsus and 1 or 2 claws; mandibles without molar part (mola); and labrum fused to head capsule.

The suborder Adephaga may be divided into 11 families, here largely following Franciscolo (1979), viz., Rhysodidae, Paussidae, Carabidae, Trachypachidae, Gyrinidae, Haliplidae, Hygrobiidae, Amphizoidae, Phreatodytidae, Noteridae and Dytiscidae. The relationship between these families is still a matter of much discussion; for references to some of the important works on this subject, see Lawrence & Newton (1982).

The families Rhysodidae, Paussidae and Trachypachidae have been referred to the Carabidae by certain authors, and these four families are often termed the "Geadephaga" (sometimes with the exclusion of Trachypachidae), as they comprise largely terrestrial species. The Paussidae are not represented in northern Europe, and the Rhysodidae and Trachypachidae only comprise one Fennoscandian species each, viz., *Rhysodes sulcatus* (F.) and *Trachypachus zetterstedti* (Gyll.). The Fennoscandian species of Trachypachidae (as a subfamily of the Carabidae) and Rhysodidae have been treated in volume 15 of this series (Lindroth, 1985-86).

The remaining families of Adephaga are all largely aquatic, and are often termed the "Hydradephaga" or "Hydrocanthares" (sometimes with the inclusion of Trachypachidae). The Phreatodytidae (by some authors referred to the Noteridae) live subterranean and are only known from Japan. Members of the Amphizoidae spend part of their life cycle in running water; they are known only from western North America and China (Kavanaugh, 1986). The other families of aquatic Adephaga all have representatives in northern Europe. The largest family, the Dytiscidae, will be treated in a future volume, while the four small families, Gyrinidae, Haliplidae,

10

Hygrobiidae and Noteridae are the subject of this volume. One of the latest attempts to reconstruct the phylogeny of the aquatic families was made by Kavanaugh (1986).

Key to North European families of Adephaga

Adults

1   Hind coxae not extending laterally to meet epipleura of the resting elytra (Figs 1, 2). Metepisterna or metepimera distinctly contiguous with the first visible abdominal sternite. Terrestrial species with well developed projecting sensorial setae at definite points of the body 2

–   Hind coxae extending laterally to meet the epipleura of the resting elytra (Figs 3, 13, 108). Metepisterna or metepimera at most touching first visible abdominal sternite for a very short section at elytral epipleuron. Mainly aquatic species, which sometimes are pubescent, but rarely have outstanding sensorial setae 3

2(1) Metasternum without transverse suture in front of hind coxae, separating an antecoxal plate from remaining part of metasternum. A large median semicircular part of abdominal sternite 2 visible between hind coxae (Fig. 1). Antennae thick, moniliform (Fig. 4) .... Rhysodidae

–   Metasternum with a transverse suture in front of hind coxae, separating an antecoxal plate from remaining part of metasternum (Fig. 2). Median portion of second abdominal sternite hardly visible or obsolete. If antennae thick, then not moniliform .... Carabidae (in part) (see Vol. 15)

3(1) Body with outstanding sensorial setae at fixed points. Terrestrial species resembling species of *Bembidion* (Carabidae) ....
.... Carabidae, Trachypachinae (see Vol. 15)

–   Body sometimes pubescent, but without outstanding sensorial setae at fixed points. Aquatic species 4

4(3) Each compound eye completely divided into a dorsal and a ventral portion (Fig. 16). Antennae very short and thick (Fig. 12) .... Gyrinidae (p.16)

–   Each compound eye not completely divided. Antennae longer 5

5(4) Parts of hind coxae produced into large flat plates concealing basal abdominal sternites (Fig. 108) .... Haliplidae (p.65)

–   Parts of hind coxae not produced into plates concealing basal abdominal sternites 6

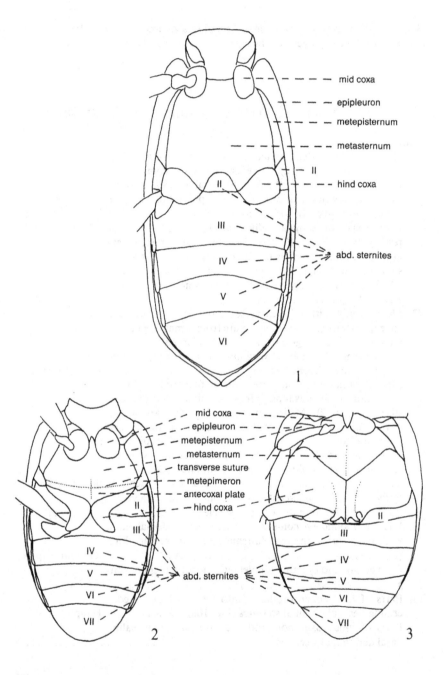

mid coxa

epipleuron

metepisternum

metasternum

II

hind coxa

abd. sternites

II

III

IV

V

VI

1

mid coxa

epipleuron

metepisternum

metasternum

transverse suture

metepimeron

antecoxal plate

hind coxa

II

III

IV

V

VI

VII

abd. sternites

II

III

IV

V

VI

VII

2

3

6(5) Metasternum with a transverse suture in front of hind coxae,
separating an antecoxal plate from remaining part of metasternum (Fig. 256). Compound eyes strongly protruding
(Fig. 255). Hind legs moved alternately when swimming . Hygrobiidae (p.136)
– Metasternum without a transverse suture. Compound eyes
not strongly protruding beyond general outline of head.
Hind legs moved simultaneously when swimming 7
7(8) Body with dorsal side strongly convex, ventral side flattened.
Hind coxae rather small, produced into plates of a characteristic shape medially (Fig. 267). Metepisterna very clearly
not reaching the mid coxal cavities (Fig. 267) . . . . Noteridae (p.143)
– Body with dorsal and ventral sides almost equally convex.
Hind coxae rather large, not forming plates of the shape
seen in Noteridae medially (Fig. 3) . . . . Dytiscidae

Larvae (partly after Thomson, 1979)
1 Larvae from aquatic habitats. Respiratory system often
closed; if open, then abdomen with less than 8 pairs of la-
teral spiracles 5
– Larvae from terrestrial habitats. Respiratory system open;
abdomen with 8 pairs of lateral spiracles 2
2(1) Urogomphi absent 3
– Urogomphi of some size and distinct shape present (Figs 5, 6) 4
3(2) Labial palpi absent. Larvae usually in wood, under bark or in
roots . . . . Rhysodidae
– Labial palpi present. Larvae associated with ants, ectopara-
sitic, or with at least one pair of hooks on dorsal side of
abdominal segment 5 . . . . Carabidae (in part)
4(2) Larvae with the following combination of characters: head
not constricted into a narrow neck posteriorly, with 6 ocelli
on each side, and without transverse grooves on the sides
between the ocelli and the posterior margin (Fig. 7). Cutting
edge of mandible flattened into a thin plate; cutting edge
and retinaculum not denticulate or crenulate (Fig. 8). Inner
lobe of maxilla absent, article 1 of outer lobe fused with
stipes (Fig. 9). Labium without a ligula (Fig. 10). Tarsi with
two equal claws of moderate length; claws without basal
tooth or other appendages . . . . Carabidae, Trachypachinae

Figs 1-3. Morphology of underside of 1: *Clinidium canaliculatum* Costa (Rhysodidae); 2: *Ago-
num dorsale* (Pont.) (Carabidae); 3: *Deronectes moestus* Fairm. (Dytiscidae). Redrawn from
Franciscolo (1979).

13

Fig. 4. Antenna of *Rhysodes sulcatus* (F.) (Rhysodidae).

-    Larvae without the above combination of characters .... Carabidae (in part)
5(1) Tarsi with one claw (Fig. 123) .... Haliplidae (p.65)
-    Tarsi with two claws (Figs 40, 259, 272, 273)........................... 6
6(5) Body with long ventral or lateral gill filaments (Figs 37, 261) 7
-    Body without long gill filaments 8
7(6) Abdomen with 8 well developed segments (segment 9 rudi-
     mentary). Gill filaments ventral (Fig. 261) .... Hygrobiidae (p.136)
-    Abdomen with 10 well developed segments. Gill filaments
     lateral (Fig. 37) .... Gyrinidae (p.16)
8(6) Mandibles stout, not channelled, with a retinaculum (Fig.
     271). Abdominal segments 1-7 of subequal shape and size
     (Fig. 270). Larvae burrowing among submerged roots of
     semi-aquatic plants................................. Noteridae (p.143)
-    Mandibles more slender, channelled, without a retinaculum
     (Fig. 11). Abdominal segments 1-7 not of subequal shape
     and size. Larvae not burrowing among submerged roots......... Dytiscidae

State of knowledge and distribution

The study of the Fennoscandian fauna of Gyrinidae, Haliplidae and Noteridae began
with Linnaeus (1758). Other important early works are largely the same as mentioned
for the Carabidae in Vol. 15 (Lindroth, 1985). In those years the small families of
aquatic Adephaga did not attract much interest among coleopterists; it was often
troublesome to bring along extra equipment for their collection in the field, and the
species are all rather small and often difficult to determine. The determination
problems caused much confusion regarding the identity and synonymy of species
mentioned by early authors. These problems were first solved by the end of the

Figs 5, 6. Posterior part of larval abdomen (dorsal view) of 5: *Trachypachus gibbsi* LeConte
(Carabidae: Trachypachinae); 6: *Pelophila borealis* (Payk.) (Carabidae: Carabinae); abdominal
segment X is omitted. Redrawn from Lindroth (1960b) and Andersen (1970).
Figs 7-10. Head and mouthparts of *Trachypachus gibbsi* LeConte (Carabidae: Trachypachinae),
second instar larva. – 7: head in dorsal view; 8: mandible; 9: maxilla; 10: labium. Redrawn from
Lindroth (1960b).
Fig. 11. Larval mandible of *Copelatus* sp. (Dytiscidae). Redrawn from Bertrand (1972).

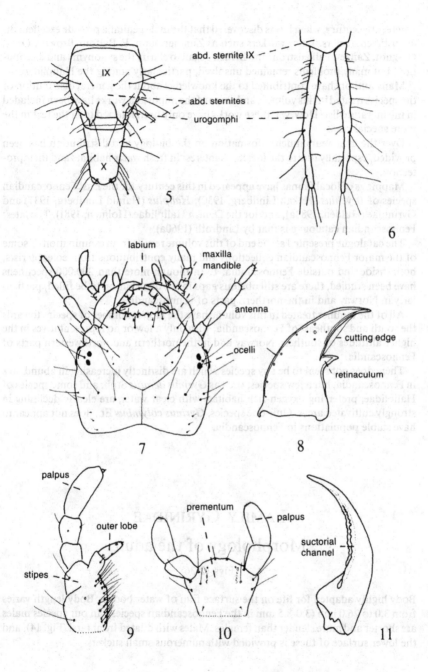

abd. sternite IX

abd. sternites

urogomphi

5

6

labium

maxilla
mandible

antenna

cutting edge

retinaculum

ocelli

7

8

palpus

prementum

palpus

outer lobe

suctorial
channel

stipes

9

10

11

nineteenth century, when it was discovered that the male genitalia provide excellent diagnostic characters. Later workers such as Zimmermann, F. Balfour-Browne, Ochs, Guignot, Zaitsev, Falkenström and Brinck began to clarify the synonymy and distribution, but many problems remained unsolved, particularly among the Haliplidae.

Many authors have contributed to the knowledge about the general distribution of the species treated in this volume, and some of the important works have been included in the literature list. However, these works have only been referred to in the text in the more special cases.

Over the later years much information on the biology and distribution has been provided, especially due to the increased interest in fresh water habitats and their protection.

Mappings of local faunas have appeared in this century for the East Fennoscandian species of *Brychius* (Håkan Lindberg, 1930), *Haliplus* (Harald Lindberg, 1937) and Gyrinidae (Huldén, 1983a), and for the Danish Haliplidae (Holmen, 1981). The latest Fennoscandian catalogue is that by Lindroth (1960a).

The catalogue presented at the end of this volume results from examination of some of the major Fennoscandian collections and many contributions from coleopterists, both inside and outside Fennoscandia. Even though more than 200.000 specimens have been studied, there are still obvious gaps in the distribution to be filled, particularly in Norway and in the northern parts of Fennoscandia.

All of the families treated in this volume have the largest number of species towards the south and southeast of Fennoscandia, with only a few or no representatives in the higher altitudes of southern Norway and in the northern and northwestern parts of Fennoscandia.

There does not seem to be any species which are distinctly increasing in abundance in Fennoscandia, but a few species, such as *Gyrinus aeratus* Stph. and some species of Haliplidae, preferring oxygen-rich habitats with clear water, are clearly declining in strongly cultivated areas. Only one species, *Gyrinus colymbus* Er., does not appear to have stable populations in Fennoscandia.

FAMILY GYRINIDAE
Morphology of the adult
(Figs 12-36)

Body highly adapted for life on the surface film of water bodies. Body length varies from 3.0 to 26.0 mm (3.0-8.5 mm in the Fennoscandian species). In our species males are shorter and more slender than females. Males with dilated front tarsi (Fig. 14), and the lower surface of these is provided with numerous small suckers.

Head (Figs 12, 13, 16)

Prognathous and fairly large (Fig. 16). Eyes completely divided into a ventral and a dorsal portion. Labrum transverse, either fairly wide and short, or narrow and strongly protruding; rounded with a marginal fringe of long hairs; punctuation, reticulation, wrinckles, or short pubescence may cover its dorsal surface. Labrum covered by the clypeus; this is transverse and sometimes very narrow. Clypeus posteriorly fixed to the frons with a distinct suture; it usually has lateral extensions, the clypealia. Frons and vertex fused without any visible sutures. Dorsal and ventral eye portions separated by the interorbital area, and sometimes partly by the antennal cavity; this is anteriorly delimited above by the frontal ridge. Ventral side of head, behind the mouthparts, consisting of a median gula and the lateral genae.

Antennae (Fig. 16)

These are short, primitively 11-segmented, but as many as four of the distal segments may be fused. Basal two segments large; second segment flattened, fringed with hairs, bearing the slightly club-shaped remaining part of the antenna. Apical segment often longer than each of the other segments in the club-shaped part of the antenna.

Mouthparts (Figs 13, 16-19)

Labium (Fig. 17) with a large lobate mentum which partly surrounds the well developed ligula. Labial palps composed of three segments, the most distal of which usually is the longest, and a distinct palpiger. Maxillae (Fig. 18) with a curved lacinia having long spines along the inner edge, a long one-segmented galea (sometimes absent), and a four-segmented palpus on a palpiger. Mandibles (Fig. 19) stout and strongly curved, the sometimes bifid apex with a vertical cutting-edge; also with a small retinaculum and often some additional teeth along the inner edge.

Thorax (Figs 12, 13)

Pronotum well developed and covering entire dorsal surface of prothorax, continuing as the hypomeron ventrally. Pronotum often with transverse depressions, a transverse, medially widely interrupted row of punctures near the anterior edge, and a mid-line. Surface may be punctured and micro-reticulate. Anterior pronotal edge usually curved, protruding in the middle, and with more or less produced anterior corners. Posterior edge slightly curved, in European species only weakly backwardly projecting in the middle, and with slightly produced posterior corners. Pronotal sides often flattened, with a more or less distinct lateral border. Prosternum transverse, laterally reaching the hypomera, with a median ridge continuing on the prosternal apophysis. The latter only weakly developed, and even absent in some species. Proepisterna well developed, reaching the front coxal cavities. Notopleural suture present.

Mesothorax fairly large. Most of its dorsal surface is covered by the elytra, excepting the scutellum, which is exposed in European species. Mesosternum large, rhomboid, bordered anteriorly; often with groups of punctures and a median furrow; posteriorly

17

attached to the immovable triangular mid coxae. Both mesepimera and mesepisterna visible from below, the former sometimes separating the latter from the epipleuron. Mesocoxae fairly large, immovably fused with the meso- and metasterna.

Metathorax large, dorsally covered by the elytra. Most of the ventral side is occupied

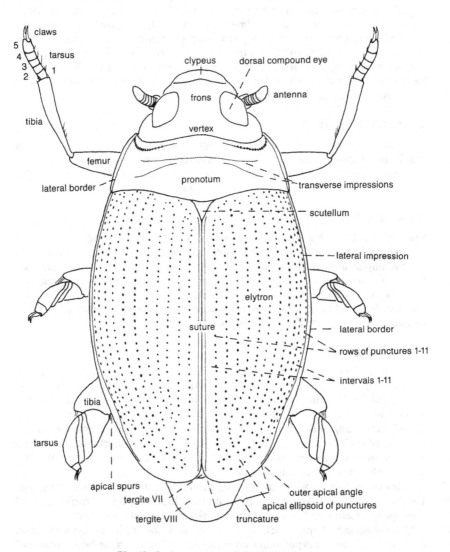

Fig. 12. *Gyrinus marinus* Gyll., dorsal view.

by the large, squarish hind coxae which reach the epipleura. Metasternum transverse, fairly small, perhaps in one North American species transversely divided by an antecoxal suture, sometimes with a median ridge. Narrow lateral parts of metasternum separated from the epipleura by the metepisterna; median part produced anteriorly as the metasternal apophysis, reaching the mesosternum and separating the mid coxae. Metepimera not visible from below. For further details on the morphology and functioning of the thorax and its appendages, see Larsén (1966).

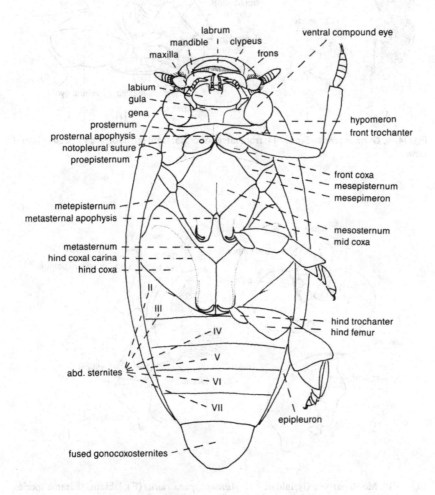

Fig. 13. *Gyrinus distinctus* Aubé, ventral view.

19

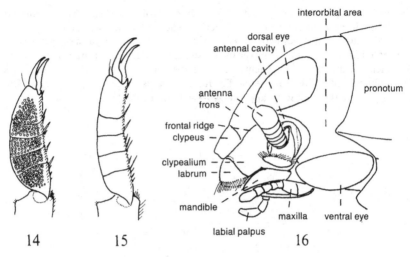

Figs 14-16. *Gyrinus aeratus* Stph. – 14: front tarsus of ♂; 15: front tarsus of ♀; 16: head in lateral view.

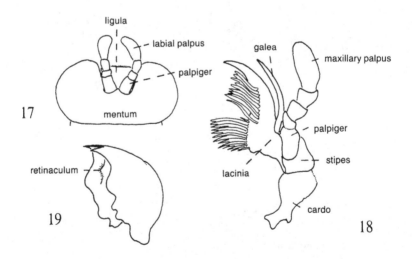

Figs 17-19. Mouthparts of Gyrinidae. – 17: *Aulonogyrus striatus* (F.), labium; 18: same species, maxilla; 19: *Gyrinus substriatus* Stph., mandible. Redrawn from Franciscolo (1979).

20

Legs (Figs 14, 15)

Front legs rather long and unmodified; front coxae subcylindrical, movable; trochanter short, triangular; femur and tibia long and slender, provided with rows of spines; tarsus with 5 fairly short, laterally compressed, segments of which the distal bears a pair of claws. Mid and hind legs short and flattened, highly adapted for swimming; coxae very large and immovably fixed to ventral thoracic sclerites; each coxa posteriorly with an apophysis below which the small triangular trochanter is inserted; femora and tibiae short, flattened, triangular. Five tarsal segments, of which the distal one bears a terminal pair of claws; they are modified to form a wide fan. Mid and hind legs provided with a number of short spines; tibiae with two slightly longer apical spurs; tibiae and tarsi furthermore furnished with rows of long lamellate swimming-hairs (derived from rows of spines).

Elytra

Free and well sclerotized, covering the abdomen, excepting the tip. In many species each elytron has 9-11 rows of fairly large punctures, in others the elytra are randomly punctured; our Gyrininae have 11 rows. Row 1 is the one closest to the suture and row 11 is nearest the side-margin; the rows are separated from the suture and each other by the intervals 1-11. The rows may be placed in shallow, strongly micro-reticulate furrows, while the intervals may be elevated; in the Gyrininae an apical ellipsoid of punctures is usually distinct. Elytra finely bordered laterally, and with a more or less well marked lateral depression along the border; this depression reaches the outer apical angle (sometimes effaced), but generally vanishes along the border between this point and the suture; the latter, apical portion of the elytra is often termed the truncature. Elytra bent down- and inwards at the sides, continuing ventrally as the elytral epipleura, each of which near the base has an oval thickening with dense spicules on the inner face. This thickening faces an oval spicule-covered dilation on the thorax which probably fixes the elytron when the insect is not flying.

Hindwings (Fig. 20)

These are membranous, hyaline and well developed, at least in our members of the family; only *Orectochilus villosus* (Müll.) has the wings slightly reduced. Venation (nomenclature following Ward (1979)) is of an Adephagan type in which the M4 originates anterior to the middle of the Oblongum Cell. A few characteristics of the gyrinid wing should be mentioned: Oblongum Cell often in shape of a parallelogram with posterior side slightly longer than anterior (different in *Gyrinus*); M4 originates variously close to, or associated with, M3; the 3R Cell larger than the SA Cell; the wedge-shaped Cell between 1A1 and 1A2 absent; subcubital binding patch of spicules absent. Wing-folding pattern also of an Adephagan type, resembling that of the Hygrobiidae, where the cubital hinge is rather far removed from the Oblongum Cell, and the folding itself mainly takes place by a spring mechanism of the veins; in most

21

other aquatic Adephaga the wings are folded by the aid of tergal or elytral binding patches.

Abdomen (Fig. 21)

Abdomen dorsally with 8 visible, partly membranous tergites (segments I-VIII), and a number of partly membranous pleurites on each side. These are, with the exception of T VIII and sometimes parts of T VII, covered by the elytra. Ventrally 6 sternites (II-VII) are exposed. According to Burmeister (1976) the distal ventral abdominal sclerite is not a true sternite, but the fused gonocoxosternites (not fused in one North American species); ventral parts of abdominal segment VIII give rise to the gonocoxosternites which are invaginated in other families of Adephaga. Lateral parts of S II-VI (laterosternites) are covered by the elytra. Sternites II-IV fused, usually with visible sutures, and furthermore immovably fixed to the hind coxae along their anterior margin. In the Orectochilinae S VII and the fused gonocoxosternites are long and rather narrow, provided with a median row of long hairs; these segments are shorter and wider in the Gyrininae. The fused gonocoxosternites are rounded apically. Tergite IX and sternites VIII-IX invaginated, more or less reduced or modified to function as parts of the reproductive system. Pleural membranes I-VI and tergal margins VII-VIII with a pair of spiracles on each segment.

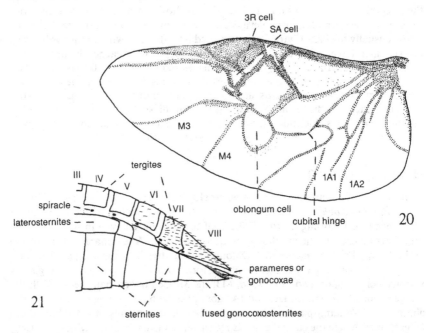

Figs 20, 21. *Gyrinus aeratus* Stph. – 20: hind wing; 21: posterior part of abdomen in lateral view.

Male genitalia (Fig. 22)

External genitalia mainly comprised of the invaginated segment IX and its appendages (Fig. 22). Tergite IX hardly distinguishable, perhaps present as two very small sclerites near the anal conus. Sternite IX modified, forming a sheath ventrally enclos-

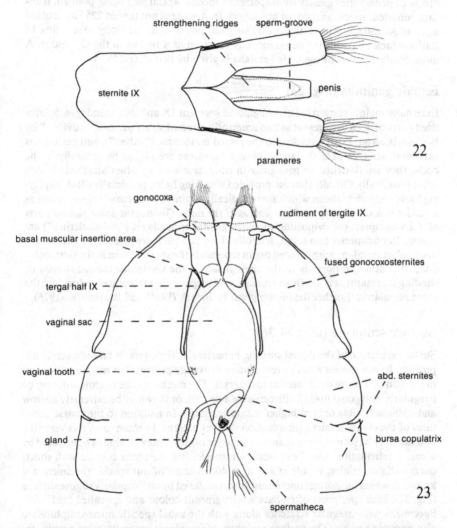

Figs 22, 23. *Gyrinus substriatus* Stph. – 22: male genital sclerites in ventral view; 23: parts of female reproductive system with genital sclerites, dorsal view. Redrawn from Brinck (1955).

ing the aedeagus at rest. Aedeagus consisting of two long lateral lobes, the parameres, partly enclosing a median phallic lobe, the penis; all three fused proximally. Aedeagus bilaterally symmetrical. Parameres flattened or triangular in transverse section, with an apical fringe of long hairs. Penis often flattened in our species, rather strongly sclerotized, with a ventral sperm-groove, and with a number of strengthening-ridges. Shape of penis varies greatly from species to species, apical part being pointed, truncate, rounded, spoon-shaped, or apically cleft. Aedeagus not turned 225° as in other aquatic Adephaga, but resting in its original position in the body. According to Balfour-Browne (1950) the aedeagus may function as a rudder in the Gyrinidae. A more detailed survey of the male genitalia is given by Brinck (1955).

Female genitalia (Fig. 23)

Externally mainly comprised of invaginated segment IX and its appendages. Sclerotized parts of T IX are present as two anteriorly directed struts (termed "valvifers" by Balfour-Browne, 1950), attached to the paired gonocoxae ("valves"), and perhaps as two small sclerites near the anal conus. Gonocoxae are placed horizontally in the body; they are flattened or triangular in transverse section, often attached to each other proximally. Distally they are provided with long hairs, proximally often displaying folds and indentations which are specifically distinct. Gonocoxae may function as a rudder in a similar way to the aedeagus of the male. Gonocoxae and attached parts of T IX comprise the ovipositor, in which genital appendages ("vulvosclerites") are absent. In our species two small, sclerotized vaginal teeth are present. A well sclerotized, often spiral- or horn-shaped organ represents the spermatheca in the Orectochilinae; it is attached dorsally to the vaginal sac. In the Gyrininae the sac-shaped or winding spermatheca is hardly sclerotised, and it opens via a receptacular duct into the bursa copulatrix. Further details are given by Brinck (1955) and Burmeister (1976).

Surface sculpture (Figs 24-36)

Surface sculpture of the dorsal surface, in particular the elytra, is taxonomically important. In our species a micro-reticulation is usually present, but magnifications of more than × 150 are often needed to detect it. The meshes of the reticulation may be irregularly polygonal (usually discernable at × 50), or they may be extremely narrow and oblique (visible only at higher magnifications). In addition to the coarse punctures of the elytra, a micro-punctuation is often present. In some cases it is very fine and can only be seen at more than × 150, in yet others it may be almost obliterated by a coarse reticulation. One Fennoscandian species has the elytra covered with short, deep, oblique strioles, visible at about × 40. In some of our species specimens are known in which the normal microsculpture is replaced by an irregularly rugose surface (Fig. 32); such specimens often have a ferrugineous colour and are called "rufinos". Specimens with irregular scratches along with the usual specific microsculpture are also known to occur. The surface sculpture of females is generally more deeply impressed than in males.

24

Dorsal pubescence is a characteristic of the subfamily Orectochilinae (Fig. 36). Most other Gyrinidae, excepting non-European species, are largely glabrous.

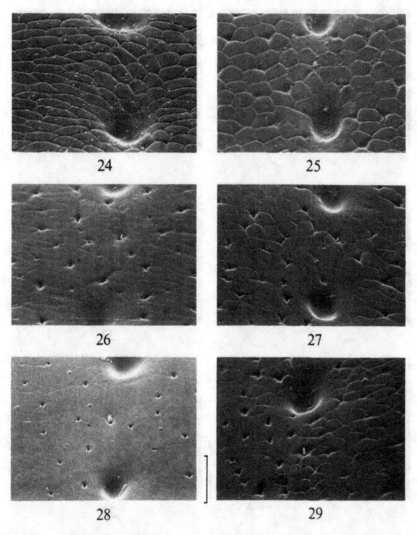

Figs 24-29. Dorsal elytral surface of *Gyrinus* spp., electroscan micrographs. – 24: *G. minutus* F.; 25: *G. opacus* Sahlbg.; 26: *G. aeratus* Stph. ♂; 27: same species ♀; 28: *G. pullatus* Zaits. ♂; 29: same species ♀. Scale: 0.02 mm.

25

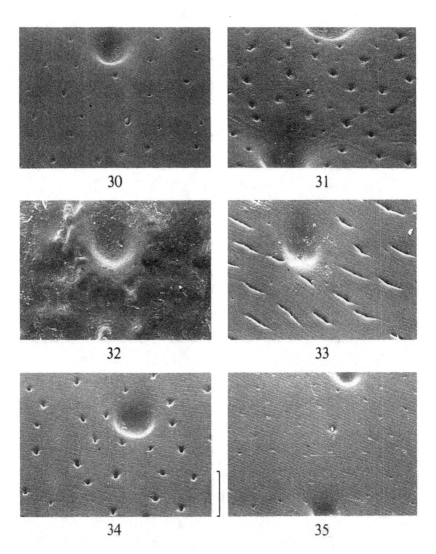

Figs 30-35. Dorsal elytral surface of *Gyrinus* spp., electroscan micrographs. – 30: *G. marinus* Gyll. ♂; 31: same species ♀; 32: same species, "rufino"-specimen; 33: *G. colymbus* Er.; 34: *G. distinctus* Aubé; 35: *G. substriatus* Stph. Scale: 0.02 mm.

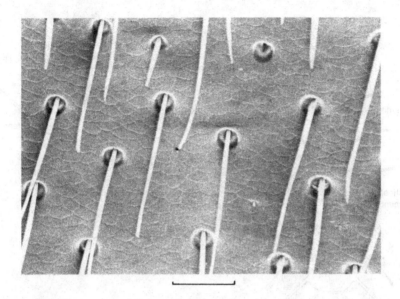

Fig. 36. Dorsal elytral surface of *Orectochilus villosus* (Müll.), electroscan micrograph. Scale: 0.04 mm.

Morphology of the larva

(Figs 37-40)

Body long, rather parallel-sided, often of a greyish colour. Head prognathous. Head capsule dorsally with an anterior fronto-clypeal sclerite which may have the front edge produced into a varying number of lobes, and two epicranial sclerites. The three sclerites are separated by distinct coronal and frontal sutures. A neck-region is absent in our species. Head on each side with a group of 6 ocelli. Labrum absent.

Antennae 4-segmented; basal segment short, other segments fairly long.

Mandibles long, curved and slender, with a suction-tube formed by a more or less closed groove. At least in some species a row of short teeth along the inner edge is present.

Maxillae with a large cardo and a somewhat smaller stipes. The 3-segmented palpus is carried by a palpiger. Lacinia short, with spines which may vary strongly with regard to number, length, and shape. Galea 2-segmented.

Labium with a short and wide base, representing mentum and submentum, distally giving rise to two long palpigers, each of which carry a long two-segmented palpus. Ligula absent.

Buccal orifice reduced, present in the narrow slit between the fronto-clypeus and the labium.

27

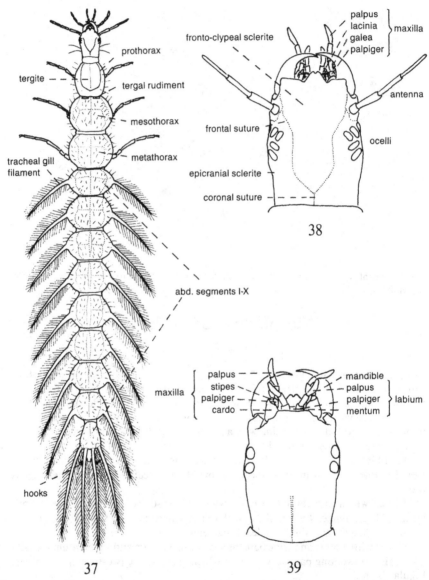

Fig. 37. Third instar larva of *Gyrinus marinus* Gyll., dorsal view. Redrawn from Schiødte (1862). Figs 38, 39. Head of larva of *Gyrinus* sp. – 38: dorsal view; 39: ventral view.

Thorax largely soft, membranous; only prothorax with a large well sclerotised dorsal tergite; small well-sclerotised tergal rudiments are present on mesothorax. Ventrally the segments bear a number of small sclerotised plates at the base of the legs, and there is often one plate transversely in front of the anterior legs. Thorax usually with a tuberculate surface and a scattered pubescence.

Legs (Fig. 40) not very long, simple, each composed of a coxa, trochanter, femur, tibia and tarsus with a small empodium and two simple terminal claws. Legs provided with a small number of spines, without swimming hairs; modified for crawling on bottom substratum or submerged vegetation.

Abdomen with 10 soft, membranous segments. Segments 1-9 of subequal shape and size, segment 10 very small. Each of the segments 1-8 carries a pair of long, haired, lateral, tracheal gill-filaments; segment 9 carries two pairs of such filaments. Segment 10 with a number of long terminal hooks (4 in our species). Abdominal segments dorsally with transverse and longitudinal furrows; their surface tuberculate with scattered pubescence. Segments 1-8 without spiracles.

Larvae of the first two instars (L I and L II) with fewer hairs than larvae of the third (and final) instar (L III) (Saxod, 1965b). For further information on the immature stages, see Bertrand (1972).

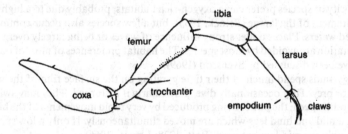

Fig. 40. Leg of larva of *Gyrinus* sp. Redrawn from Bertrand (1972).

Zoogeography

The Gyrinidae is the second largest of the families of aquatic Adephaga with more than 900 described species (Franciscolo, 1979). The family occurs in all faunal regions, with the majority of species in the tropics (no less than about 80% of the species are tropical according to Franciscolo (l.c.)).

The Holarctic region is totally dominated by the almost world-wide distributed genus *Gyrinus*. Only rather few species of the subfamilies Orectochilinae and Enhydrinae, which are dominant in the tropics, occur in the Palearctic and Nearctic regions (*Dineutus*: both Pale- and Nearctic; *Orectochilus*: Palearctic; *Gyretes*: Nearctic).

Two species, *Gyrinus minutus* and *G. opacus* are holarctic, occurring in parts of N. America and with a Eurosiberian distribution in the Palearctic region; only *G. minutus* has been recorded (but rarely) as far south as the Mediterranean area.

Eurosiberian species not or almost not penetrating into the Mediterranean area are *Gyrinus aeratus, G. pullatus* and *G. natator*. Also Eurosiberian, but penetrating deeper into the Mediterranean area, are the following species: *Gyrinus marinus, G. distinctus, G. caspius, G. paykulli, Aulonogyrus concinnus* and *Orectochilus villosus*.

A West Palearctic element comprises *Gyrinus colymbus, G. suffriani, G. substriatus, G. urinator* and *Aulonogyrus striatus*. These species do not occur in the East Palearctic, although a few may reach east of the Ural mountains and to Afghanistan. They are widely distributed in the Mediterranean area, often also in N. Africa; in our area they mainly have a southern or south-eastern distribution.

Bionomics

Our species of Gyrinidae inhabit both fresh and brackish waters. They may be met with near the shores of lakes, in ponds and marshes, and in slowly running streams. The majority of species prefer clean oxygen-rich habitats, probably due to a high oxygen requirement of the larvae (Brinck, 1955), but a few species also occur commonly in polluted waters. Places under strong influence of waves or being largely overgrown with vegetation are avoided by most species. The habitat preferences of most of our gyrinids have been discussed by Svensson (1969).

Adult gyrinids spend much of their time gyrating on the surface film of the water hunting for prey. They occasionally dive, particularly if disturbed. They may swim at a considerable speed, the power being produced by very rapid movements of the highly specialized mid and hind legs which are moved simultaneously. If only a low speed is required, only the mid legs are used (Bott, 1928; Larsén, 1966).

An interesting phenomenon often observed is the occurrence of "schools" that may consist of several hundreds of individuals, sometimes including several species. Our species of Gyrinidae seem to be more or less active in both day and night. An exception may be *Orectochilus villosus,* which is generally inactive during most of the day, when it hides ashore under stones and the like.

The adults are carnivorous, attacking insects which have been trapped by the water surface.

Below the surface an air supply is carried along under the elytra. This supply must be renewed from time to time, but some gas-exchange may take place between the water and an air-bubble protruding from below the tip of the elytra (physical gill).

Most of our species of Gyrininae have been observed flying, and specimens examined all have fully developed wings and flight musculature. Species of *Gyrinus* are often attracted to light. *Orectochilus villosus* obviously does not fly, and also shows reductions of the flight musculature (Larsén, 1966).

The eggs of Gyrininae are laid in the spring or summer; those of *O. villosus* have been found in August (Berg, 1948). The eggs are laid in rows or groups under water, glued to the surface of vegetation, to stones or the like. The egg (Fig. 41) is elongate, of white or yellowish colour. In species of *Gyrinus* it is 1.0-1.6 mm long, and is provided with reticulate ornamentation and a longitudinal adhesive band (Saxod, 1964; 1965b). The incubation time is from 8 to 12 days at about 20°C (Saxod, 1965a).

The larvae usually inhabit the same habitat as the adults. They crawl about on the vegetation or the bottom substratum, to which they cling by help of the hooks on the terminal abdominal segment. They are also capable of swimming by up- and downward undulations of the body.

Gyrinid larvae are carnivorous, preying on various small invertebrates, chiefly probably Oligochaeta and chironomid larvae which are pulled out from their burrows with the mandibles (Balfour-Browne, 1950; Bertrand, 1972). Whether the food is ingested through the mandibles is not quite known.

The larvae do not need to come to the surface to respire, as gas-exchange takes place through the thin cuticle, mainly probably on the tracheal gills.

The larvae of Fennoscandian Gyrininae pupate in the summer or autumn, and probably never hibernate. Larval development of some species has been found to last 15-35 days at about 20°C (Saxod, 1965a). The larva of *O. villosus* hibernates in the water and pupates the following spring.

Before pupating the *Gyrinus*-larva collects a large lump of debris (or other small particles) from near the water surface and carries it on its back to a suitable place above

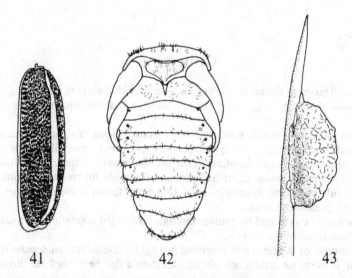

41   42   43

Figs 41, 42. *Gyrinus substriatus* Stph. – 41: egg; 42: pupa in dorsal view. Redrawn from Saxod (1964, 1965b).
Fig. 43. Pupal cocoon of *Gyrinus* sp., on a leaf of *Juncus*.

31

the water (often the stem of a plant); here the lump is moulded by the mandibles and attached to the substratum. The larva now bores itself into the lump and converts it into a pupal chamber (Fig. 43) where it rests a few days before pupation (Balfour-Browne, 1950). Somewhat similar cocoons, built by various types of material, are produced by members of our other genera (Bertrand, 1972).

The pupa (Fig. 42) is whitish, slightly shorter than the adult. It is easily distinguished from that of other aquatic Coleoptera on the large divided eyes, the short mid and hind legs, and the absence of urogomphi (urogomphi are also absent in the haliplid pupa). A key to pupae of the Fennoscandian subfamilies is provided by Bertrand (1972).

The pupa rests on its posterior end. Its dorsal side carries setae to separate it from the inner wall of the cocoon. In some species the time spent by larva, pupa and adult in the cocoon has been stated to be 8-14 days at about 20°C (Saxod, 1965a). The development takes longer time for larger species than for small.

The young adults of our Gyrininae hibernate under water, reproducing in the following year; according to Saxod (1965a) only one generation is found per year. Adults of *O. villosus* have not been found from mid autumn to mid spring, and it seems likely that they die after oviposition.

Species of *Gyrinus* are hosts of ectoparasitic fungi belonging to the Laboulbeniales. *Laboulbenia fennica* Huldén has been found on *G. minutus, G. aeratus, G. pullatus, G. marinus, G. distinctus, G. natator, G. substriatus,* and *G. paykulli* (Huldén, 1983b). Species of Laboulbeniales also occur on *G. urinator* and *G. colymbus.*

Collecting

Adults and larvae of Gyrinidae are best collected by using a strong entomological net. The handle should be long, and the bag rather wide and deep with a mesh size of 1-2 mm.

Adults are usually caught selectively when seen on the surface, but even more random hauls under the surface may well produce specimens, especially during periods with cold weather. A rapid handling of the net is necessary, as the insects often move with a considerable speed. Light-traps are useful at night for collecting flying specimens. Adults of *Orectochilus villosus* are sometimes found by turning stones on the shores of streams and lakes.

Larvae can be obtained by hauling through submerged vegetation or through the uppermost layers of the bottom.

The content of the net is best examined in a light-coloured tray with water over the bottom. If too much mud is brought up, this should first be washed out through the net.

Pupal cocoons can be searched for on vertical faces above the surface of the water, such as vegetation, poles, etc., protruding out of the water.

Rearing

The rearing of species *ex ovo* is likely to produce much new information on the Gyrinidae, but strangely enough the method has not been practised much. Saxod (1965a) was successful in rearing species of *Gyrinus,* and his method is described below.

Adults are difficult to keep in aquaria for longer periods, but pregnant females may be kept long enough for them to deposit their eggs.

Each young larva is transferred to a water-filled glass vial containing water plants and a pebble on the bottom. The larva is fed with larvae of Chironomidae, and the water should be changed two or three times a week.

When the third instar larva becomes dark brown, it is transferred to a tiny water-filled dish with a pebble on the bottom, placed in the moist sand covering the bottom of a plastic box. The inner sides of the box should be lined with flat stones on whose vertical faces the pupal cocoon may be built. Before pupation the box is covered by a transparent lid. After pupation this is replaced by a net, not allowing the adult to escape. The sand should be kept moist, but not wet.

Killing and mounting

The simplest way to conserve Gyrinidae is to keep the specimens in 80% ethanol in glass tubes. This method stiffens the specimens somewhat, but it is often used in surveys where a large amount of material is conserved for later examination.

Specimens that are to be mounted dry should be killed by ethyl acetate. A few drops of this substance in a glass tube keep the specimens soft and relaxed for long periods, ready for mounting and dissection. The specimen is usually mounted on a rectangular piece of cardboard of a larger size than the insect; a water soluble glue is preferable for mounting in this way. Other ways of mounting may better allow an examination of the ventral side: the insect may be pinned with a minute pin through the basal part of the right elytron, or it may be glued to the tip of a small triangular piece of cardboard.

Examination of the genitalia is often necessary for positive identification. The genitalia can be pulled out through the abdominal tip of specimens that have not stiffened; a fine forceps and some insect pins are needed for the dissection. The genital sclerites can be cleaned from ligaments and muscles in a drop of water under high magnification, and then glued to a piece of cardboard on the same pin as the specimen. If a more detailed examination of external or internal genitalia is needed, the following procedure is recommended: the extruded genitalia are placed for a couple of days in cold caustic potash (10% KOH), afterwards carefully washed in water and then in absolute alcohol, and finally cleared for at least a day in a drop of clove-oil. They can now be transferred to a permanent slide in canada-balsam or euparal between glasses; small transparent pieces of plastic attached to the same pin as the specimen can be used instead of glasses. Enough canada-balsam or euparal to avoid distortion of the curved penis should be used.

33

Nomenclature

Names of subfamilies and tribes follow Brinck (1955).

The generic names used for the Gyrinidae are those which have generally been accepted during this century, though sometimes attributed to incorrect original authors.

Subgeneric names are used for all genera, following many recent authors. In the genera *Gyrinus* and *Orectochilus* this subdivision is not based on phylogenetic works, and the value of their subgenera may be questioned as done by Ochs (1967) for *Gyrinus*.

The specific names of the Fennoscandian species are in accordance with Silfverberg (1979) and Biström & Silfverberg (1983). Ochs (1967) is largely followed regarding the synonymy of species of *Gyrinus*, Brinck (1955) for the species of *Aulonogyrus* and Zaitsev (1953) for the species of *Orectochilus*. A number of names, previously often cited as homonyms, have, however, proven not to be valid names.

Synonyms which cannot be assigned to a species with certainty, as well as synonyms not important to the north European fauna, are generally omitted.

Key to subfamilies and genera of Gyrinidae

Adults

1   Most of dorsal surface densely pubescent (Fig. 36). Elytral
    punctures not arranged in rows. Length: 5.0-8.0 mm. (Orec-
    tochilinae) .... *Orectochilus* Dejean (p.63)
–   Dorsal surface largely glabrous. Elytral punctures arranged
    in rows (Gyrininae) 2
2(1) Pronotum and elytra with a wide yellow margin laterally.
    Elytral rows of punctures follow shallow, strongly micro-
    reticulated furrows which are particularly prominent later-
    ally (Figs 98, 99). Length: 5.1-8.0 mm .... *Aulonogyrus* Motschulsky (p.58)
–   Pronotum and elytra without a yellow margin. Elytral rows
    of punctures sometimes sulcate laterally, but not following
    well delimited, strongly micro-reticulated furrows. Length:
    3.0-8.5 mm .... *Gyrinus* Müller (p.35)

Third instar larvae

1   Anterior fronto-clypeal margin truncate in the middle (Fig.
    44). (Orectochilinae) .... *Orectochilus* Dejean (p.63)
–   Anterior fronto-clypeal margin produced medially into two
    or four lobes (Figs 38, 45). (Gyrininae)............................... 2
2(1) Anterior fronto-clypeal margin produced medially into four
    lobes (Fig. 45)...................... *Aulonogyrus* Motschulsky (p.58)

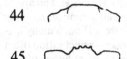

44

45

Figs 44, 45. Anterior margin of fronto-clypeal sclerite in larva of 44: *Orectochilus villosus* (Müll.) and 45: *Aulonogyrus striatus* (F.).

– Anterior fronto-clypeal margin produced medially into two lobes (Fig. 38) .... *Gyrinus* Müller (p.35)

SUBFAMILY GYRININAE

Dorsal side largely glabrous (except in *Heterogyrus milloti* Legr. from Madagascar). Maxilla with a galea (Fig. 18). Scutellum not hidden by the elytra or pronotum. Elytra with punctures arranged in 9-11 rows. Metasternum not divided by a transverse suture. Gonocoxosternites fused, semicircular in outline. Sternite VII and the fused gonocoxosternites without a longitudinal row of hairs in the middle.

Brinck (1955) divided the subfamily into two tribes, viz. Heterogyrini, represented by a single species in Madagascar, and Gyrinini, represented in most parts of the world.

TRIBE GYRININI

Dorsal side largely glabrous. Clypealia large (Fig. 16). Elytra usually with 11 distinct rows of punctures. Lateral portions of metasternum with short lateral side (Fig. 13). Posterior part of hind coxa distinctly larger than the anterior part (anterior to the hind coxal carina). Male front tarsi broadened, somewhat oval, with comparatively small segment 5 (Fig. 14).

This tribe comprises 3 non-extinct genera (Brinck, 1955): *Gyrinus* and *Aulonogyrus,* both occurring in northern Europe, and *Metagyrinus,* known only from Asia.

Genus *Gyrinus* Müller, 1764

Gyrinus Müller, 1764, Fauna Ins. Fridrichsdalina: XVII (see note).
Type species: *Dytiscus natator* Linnaeus, 1758 [= *Gyrinus (Gyrinus) natator* (Linnaeus, 1758)], by subsequent designation (Latreille, 1810).

Small or medium-sized whirligigs. Body oval or elongate, often fairly depressed. Dorsal side of the European species black or bluish black (rarely reddish), usually with a bronze sheen laterally, and sometimes with bronze stripes on the elytra. Labrum short, without longitudinal furrows. Dorsal eye anterior to ventral eye, projecting beyond antennal base. Antenna 9-segmented. Pronotum with weak transverse impressions or grooves (Fig. 12). Elytra with isolated round punctures forming 11 rows and an apical ellipsoid. The rows may be slightly sulcate, but never follow well delimited, strongly micro-reticulated furrows. Elytra with reticulate micro-sculpture, although many species appear smooth at magnifications below × 150. Dorsal micro-punctuation often present, normally consisting of small rounded punctures or short oblique strioles. Specimens, often reddish (so-called "rufinos"), which have the usual sculpture greatly distorted by minute wrinkles or scratches, occur among many species (Fig. 32). Tergite VII at most weakly trilobed behind.

Gyrinus comprises about 140 species which are distributed throughout most parts of the world (Franciscolo, 1979). The European species belong to two subgenera, viz., the Holarctic *Gyrinulus* Zaitsev and the almost world-wide distributed *Gyrinus* s. str.

Members of the genus occur in both running and stagnant water where the surface is not covered by too much vegetation. Many of the European species are known to fly, and are often attracted to light. The ecology of many of our species has been treated by Svensson (1969).

Notes on synonymy. Geoffroy (1762) first described the genus *Gyrinus,* and listed as the only species *natator* Linnaeus, 1758. However, his work is not recognized by the International Commission on Zoological Nomenclature, because he did not consistently use the binomial nomenclature. Müller (1764) was next to use the generic name *Gyrinus,* attributing it to Geoffroy, and his genus clearly fits a gyrinid. Müller (l.c.) did not associate any species with the name. The *Dytiscus natator* mentioned later in the same work is, however, no doubt a species of *Gyrinus.* Linnaeus (1767) later assigned his *Dytiscus natator* Linnaeus, 1758, and also a North American species, to *Gyrinus.*

The genus *Gyrinus* has been attributed to all of the above mentioned authors in various publications. However, it seems that Müller was the first author to use the name in a recognized work, and it should therefore be attributed to him (Silfverberg, 1978).

Linnaeus (1767) was the first author to assign nominal species to the genus in a recognized work, and thus defining the genus. Latreille (1810) subsequently designated *Dytiscus natator* Linnaeus, 1758 [= *Gyrinus (Gyrinus) natator* (Linnaeus, 1758)] as the type species of *Gyrinus* Müller, 1764.

Due to the intraspecific variation of some characters used in the following key, the determination should, if possible, be confirmed by an examination of the genitalia. Special attention should be paid to the shape of the distal part of the penis, and to the basal folds and projections of the gonocoxae. The length of the specimens is measured from the anterior margin of the head to the tip of the elytra.

Key to species of *Gyrinus*

1   Entire length of mesosternum with a deep median groove
      (Fig. 46). Scutellum usually with a distinct median tubercle
      (Fig. 48). Length: 3.0-4.7 mm. (sg. *Gyrinulus* Zaitsev) . . . .
      . . . . 1. *minutus* Fabricius
–   Mesosternum with a fine median groove (suture) in the po-

sterior half; sometimes it is weakly sulcate anteriorly (Fig. 47). Scutellum without a distinct ridge or tubercle (sg. *Gyrinus* s. str.) 2

2(1)   Claws of mid and hind tarsi black, at least in the distal half. Underside, excepting legs, usually completely black with a metallic sheen 3

–   Claws of mid and hind tarsi yellow or reddish. Colour of underside varies, usually without a metallic sheen 5

3(2)   Lateral impression of elytra fairly wide, distinctly widened before tapering towards the rounded, effaced outer apical angle of the elytron (Fig. 53). Body usually widest behind the middle. Male: penis with pointed apex (Fig. 66). Female: basal edge of gonocoxa slightly sinuate, but without indentations (Fig. 79). Length: 4.5-8.0 mm ..... 5. *marinus* Gyllenhal

–   Lateral impression of elytra narrow, hardly widened before reaching the outer apical angle of the elytron which is usually distinct (Figs 51, 52). Body widest at the middle. Male: penis with a rounded or truncate apex (Figs 64, 65). Female: inner part of basal edge of each gonocoxa with an indentation (Figs 77, 78) 4

4(3)   Elytra slightly elevated between the inner rows of punctures, especially posteriorly. Highest point of the body usually in front of middle (Fig. 88). Tergite VIII almost semicircular behind (Fig. 90). Outer distal angle of front tibia somewhat protruding (Fig. 92). Male: the narrow

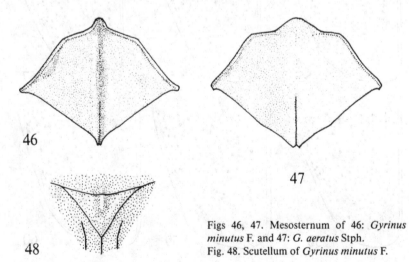

46

47

48

Figs 46, 47. Mesosternum of 46: *Gyrinus minutus* F. and 47: *G. aeratus* Stph.
Fig. 48. Scutellum of *Gyrinus minutus* F.

37

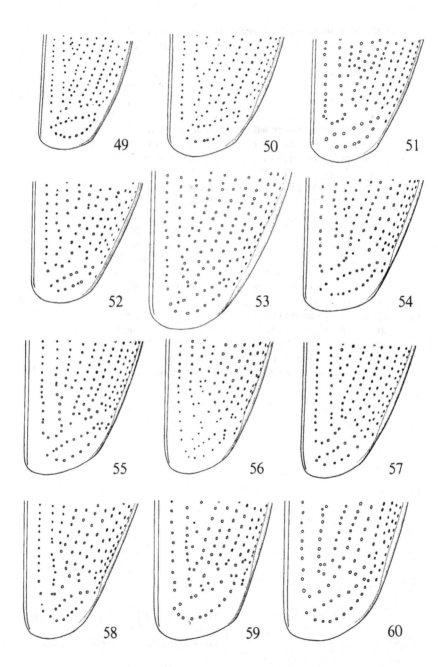

49 50 51
52 53 54
55 56 57
58 59 60

38

distal part of penis about one third of the length of penis
(Fig. 64). Female: the basal muscular insertion area on
gonocoxa of a characteristic shape (Fig. 77). Length: 4.4-
6.3 mm .... 3. *aeratus* Stephens

–   Elytra almost even between the inner rows of punctures.
Highest point of body usually at the middle (Fig. 89).
Tergite VIII more widely rounded behind (Fig. 91). Outer
distal angle of front tibia hardly protruding (Fig. 93).
Male: the narrow distal part of penis about one fourth of
the length of penis (Fig. 65). Female: the basal muscular
insertion area on gonocoxa of a characteristic shape (Fig.
78). Length: 4.5-6.0 mm .... 4. *pullatus* Zaitsev

5(2) Dorsal side with a polygonal micro-reticulation which is
often very strong (discernable at about × 40), thus giving
the species a dull appearance (Fig. 25). Ventral side, excepting legs, usually completely black. Length: 4.7-7.5 mm ....
.... 2. *opacus* Sahlberg

–   Dorsal side without a polygonal micro-reticulation, normally appearing smooth; narrow, oblique meshes may be
discernable at high magnification, × 150 (Figs 33-35).
Colour of ventral side variable 6

6(5) Elytra with stripes of bronze reflections following the
rows of punctures. The innermost rows of punctures obsolete in the middle (Fig. 61). Ventral side reddish or
yellowish, sometimes darkened. Length: 4.0-7.8 mm ...... *urinator* Illiger

–   Elytra often bronzed laterally, but without stripes following the rows of punctures. Normal colour of ventral side
black, but often with the hypomera, epipleura, mesosternum and the fused gonocoxosternites brown or red-
dish 7

7(6) Surface of elytra densely covered with short, deep, oblique
strioles (Fig. 33). Length: 5.0-7.0 mm .... 6. *colymbus* Erichson

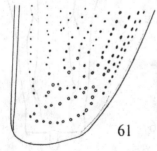

61

Figs 49-61. Apical part of right elytron of *Gyrinus* spp. –
49: *G. minutus* F.; 50: *G. opacus* Sahlbg.; 51: *G. aeratus*
Stph.; 52: *G. pullatus* Zaits.; 53: *G. marinus* Gyll.; 54: *G.
colymbus* Er.; 55: *G. distinctus* Aubé; 56: *G. suffriani*
Scriba; 57: *G. natator* (L.); 58: *G. substriatus* Stph.; 59: *G.
caspius* Mén.; 60: *G. paykulli* Ochs; 61: *G. urinator* Ill.

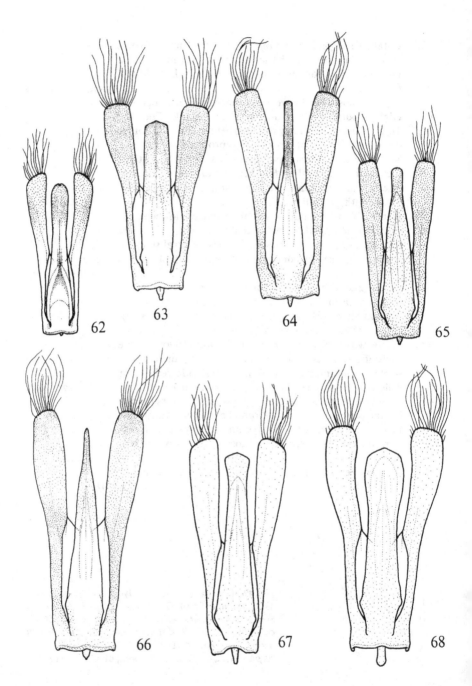

62

63

64

65

66

67

68

40

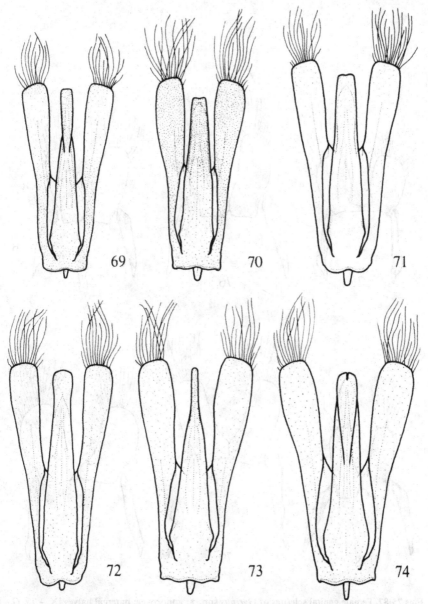

Figs 62-74. Male genital sclerites of *Gyrinus* spp., dorsal view. – 62: *G. minutus* F.; 63: *G. opacus* Sahlbg.; 64: *G. aeratus* Stph.; 65: *G. pullatus* Zaits.; 66: *G. marinus* Gyll.; 67: *G. colymbus* Er.; 68: *G. distinctus* Aubé; 69: *G. suffriani* Scriba; 70: *G. natator* (L.); 71: *G. substriatus* Stph.; 72: *G. caspius* Mén.; 73: *G. paykulli* Ochs; 74: *G. urinator* Ill.

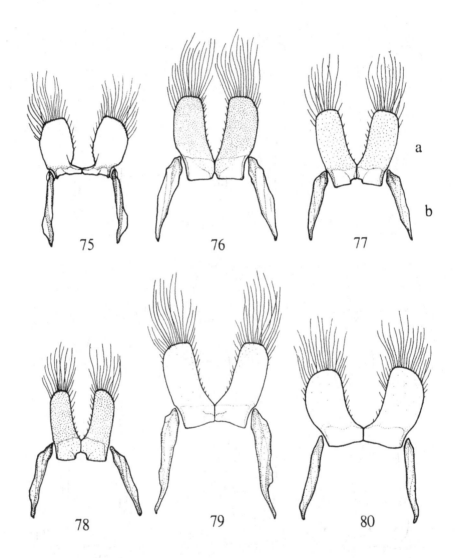

Figs 75-87. Female genital sclerites of *Gyrinus* spp., a: gonocoxae; b: tergal halves IX. – 75: *G. minutus* F.; 76: *G. opacus* Sahlbg.; 77: *G. aeratus* Stph.; 78: *G. pullatus* Zaits.; 79: *G. marinus* Gyll.; 80: *G. colymbus* Er.; 81: *G. distinctus* Aubé; 81: *G. suffriani* Scriba; 83: *G. natator* (L.); 84: *G. substriatus* Stph.; 85: *G. caspius* Mén.; 86: *G. paykulli* Er.; 87: *G. urinator* Ill.

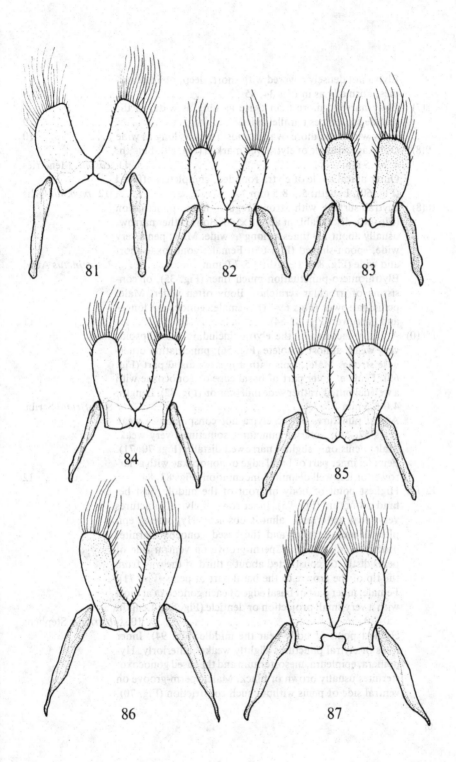

81　82　83

84　85

86　87

–   Elytra not densely covered with short, deep, oblique stri-
        oles, normally as in Figs 34, 35 8
8(7)   Body narrow, more than twice as long as wide; in the
        middle sometimes parallel-sided 9
–   Body wider and more oval, at most twice as long as wide 10
9(8)   Outer apical angle of elytra well marked (Fig. 59). Length:
        4.9-7.5 mm .... 11. *caspius* Ménétriés
–   Outer apical angle of elytra rounded, completely effaced
        (Fig. 60). Length: 5.5-8.5 mm .... 12. *paykulli* Ochs
10(8)   Elytral surface with strong regular micro-punctuation
        (Fig. 34), discernable at about × 40. Body rather narrow,
        usually about 1.9 times as long as wide. Male: penis very
        wide, spoon-shaped (Fig. 68). Female: gonocoxae short
        and wide (Fig. 81). Length: 4.5-7.0 mm .... 7. *distinctus* Aubé
–   Elytral micro-punctuation much finer (Fig. 35), or con-
        sisting of irregular scratches. Body often wider. Male:
        penis narrower (Figs 69-71). Female: gonocoxae longer
        and narrower (Figs 82-84) 11
11(10) Apical punctures of the elytra, including the ellipsoid,
        very weak, almost obsolete (Fig. 56); punctuation other-
        wise strong. Male: penis with a narrow distal part (Fig.
        69). Female: inner part of basal edge of gonocoxae with
        a well delimited, rather wide indentation (Fig. 82). Length:
        4.0-6.5 mm .... 8. *suffriani* Scriba
–   Apical punctures of the elytra not conspicuously weak
        (Figs 57, 58); anterior punctures sometimes very weak.
        Male: penis only slightly narrowed distally (Figs 70, 71).
        Female: inner part of basal edge of gonocoxae with a nar-
        rower or less well delimited indentation (Figs 83, 84) 12
12(11) Highest point of body in front of the middle, just be-
        hind pronotum (Fig. 95). Inner rows of elytral punctures
        very weak anteriorly, almost obsolete. Hypomera, epi-
        pleura, mesosternum and the fused gonocoxosternites
        usually reddish. Male: sperm-groove on ventral side of
        penis distinctly constricted about a third of the way from
        the tip of the groove to the basal part of penis (Fig. 71).
        Female: inner part of basal edge of each gonocoxa at most
        with a very small projection or denticle (Fig. 84). Length:
        5.0-7.0 mm .... 10. *substriatus* Stephens
–   Highest point of body near the middle (Fig. 94). Inner
        rows of elytral punctures slightly weaker anteriorly. Hy-
        pomera, epipleura, mesosternum and the fused gonocoxo-
        sternites usually brown or black. Male: sperm-groove on
        ventral side of penis without such constriction (Fig. 70).

Female: inner part of basal edge of each gonocoxa with a
rather large projection or denticle (Fig. 83). Length: 4.5-
6.1 mm .... 9. *natator* (Linnaeus)

Subgenus *Gyrinulus* Zaitsev, 1908

Gyrinus sg. *Gyrinulus* Zaitsev, 1908, Russk. ent. Obozr. 7: 238.
   Type species: *Gyrinus minutus* Fabricius, 1798 [= *Gyrinus (Gyrinulus) minutus*
   Fabricius, 1798], by monotypy.
Gyrinus sg. *Gyradelphus* Gozis, 1915, Miscnea ent. 23: 8.
   Type species: *Gyrinus minutus* Fabricius, 1798 [= *Gyrinus (Gyrinulus) minutus*
   Fabricius, 1798], by monotypy.

Entire length of mesosternum with a deep median groove. Scutellum usually with a
distinct median ridge or tubercle.
   At present this subgenus contains only the Holarctic species *minutus* and another
N. American species which is sometimes considered conspecific with *minutus*. According to Ochs (1967), certain S. American species of *Gyrinus* also have a median
ridge on the scutellum, and the present, generally used subdivision of the genus is
therefore not quite satisfactory.

1. **Gyrinus (Gyrinulus) minutus** Fabricius, 1798
   Figs 24, 46, 48, 49, 62, 75.

Gyrinus minutus Fabricius, 1798, Ent. Syst. Suppl.: 65.
Gyrinus Kirbii Marsham, 1802, Ent. Brit. 1: 100.
Gyrinus bicolor, sensu auctt.; ? misident., ? *nec* Fabricius, 1787; see note under *G.*
   paykulli.

3.0-4.7 mm. Narrowly oval. Dorsal side black, rarely reddish, with a bronze sheen
laterally. Ventral side yellowish, often darkened. Legs yellowish. Head, pronotum and
elytra with strong polygonal micro-reticulation, discernable at low magnification, ×
40 (Fig. 24); without micro-punctuation. Scutellum usually with a distinct ridge or
tubercle (Fig. 48). Inner rows of elytral punctures a little weaker than outer rows; apical
ellipsoid of punctures well marked; lateral impression narrow, without any dilation in
front of the outer apical angle (Fig. 49); outer apical angle prominent; meshes of
elytral micro-reticulation slightly transverse. Mesosternum for whole length with a
deep median groove (Fig. 46). Male: penis rather broadly rounded apically with a
small terminal notch (Fig. 62). Female: gonocoxa short and wide, with most of the inner edge convex (Fig. 75).

   Distribution. Rare in the eastern part of Denmark, more common in the west; not
recorded from NWZ. – Very widespread and generally common in the rest of Fennoscandia, though not recorded from a few districts in Norway and Finland. – Most
of Europe, north to Britain, Fennoscandia and the USSR, south to northern parts of

Spain, Italy and the Balkans; widespread in the USSR, Siberia, Mongolia, Manchuria; N. America.

Biology. Mainly found in larger stagnant bodies of clear or dystrophic water, but may occur in a rather wide spectrum of habitats (Svensson, 1969). Adults seem to prefer calm, open parts of the surface, where the vegetation is not very dense. Also recorded from above the timber line (Huldén, 1983a). Oviposition has been observed in late spring, and under laboratory conditions development into adult has been completed in about 34 days (Saxod, 1965a). Saxod (1965b) has described the larva and the pupa. The adult hibernates below the surface, clinging to submerged vegetation. This species is known to fly (Jackson, 1973).

Subgenus *Gyrinus* s. str.

Gyrinus sg. *Gyrinus* s. str.

Posterior half of mesosternum with a fine median groove (suture); sometimes weakly sulcate anteriorly. Scutellum in the European species without a distinct ridge or tubercle.

This almost world-wide distributed subgenus probably contains more than a hundred species. It comprises most species of Gyrinidae in the Holarctic area.

2. *Gyrinus (Gyrinus) opacus* Sahlberg, 1819
Figs 25, 50, 63, 76.

Gyrinus opacus Sahlberg, 1819, Ins. Fenn. 1: 47.
Gyrinus indicus Aubé, 1838, *in* Dejean: Spec. gén. Col. Dej. 6: 689.
Gyrinus opacus var. *lecontei* Omer-Cooper, 1930, Entomologist's mon. Mag. 66: 68; preocc., *nec* Fall, 1921.
Gyrinus opacus var. *blairi* Omer-Cooper, 1931, Entomologist's mon. Mag. 67: 239; replacement name.

4.7-7.5 mm. Oval, body widest at or slightly in front of the middle. Dorsal side black, usually with a bronze sheen laterally. Ventral side black, hypomera, epipleura, mesosternum and the fused gonocoxosternites rarely lighter. Legs yellowish, often darkened. Head, pronotum and elytra with strong polygonal micro-reticulation (rarely obliterated), discernable at low magnification, × 40; meshes isodiametric (Fig. 25), slightly oblique towards the sides. Micro-punctuation strong on pronotum, weaker on head and elytra, sometimes disappearing. Elytral rows of punctures generally rather weak, inner rows somewhat weaker than outer rows; apical ellipsoid of punctures well marked; lateral impression narrow, not or only slightly dilated before reaching the outer apical angle (Fig. 50), which is only very weakly marked. Mesosternum with a narrow groove only in the posterior half. Male: penis wide, almost parallel-sided, with

46

apex forming a blunt angle (Fig. 63). Female: gonocoxa almost as in *marinus*, with the basal edge almost without indentations (Fig. 76).

Distribution. Not in Denmark. – In most of Sweden, excepting the southwest. The previous record from Sm. (Lindroth, 1960a) is omitted, as the specimens examined proved to belong to other species. – Norway: widespread, but mainly found at higher elevations; not recorded from a number of coastal districts. – Finland: widely distributed, common in the north, and in the south mainly recorded from the archipelago; not recorded from Ka. – Adjacent parts of the USSR: known from all districts. – Iceland, Scotland, northern parts of the USSR, Mongolia; N. America, Greenland.

Biology. Mainly in stagnant water, especially peatholes and marshes. In the north it often occurs above the timber line (Huldén, 1983a). Not much is known about its life-cycle; Böcher (in press) provides a figure of a third instar larva from Greenland, collected at the beginning of August. The species is known to fly (Lundberg & Nilsson, 1978).

3. **Gyrinus (Gyrinus) aeratus** Stephens, 1832
   Figs 14-16, 20, 21, 26, 27, 47, 51, 64, 77, 88, 90, 92.

Gyrinus aeratus Stephens, 1832, Ill. Brit. Ent., Mandib. 5: 395.
Gyrinus marinus var. *thomsoni* Zaitsev, 1908, Russk. ent. Obozr. 7: 122.
Gyrinus edwardsi Sharp, 1914, Entomologist's mon. Mag. 50: 137.
Gyrinus Thomsoni ab. *zimmermanni* Franck, 1932, Ent. Bl. Biol. Syst. Käfer 28: 130.
Gyrinus opacus, sensu auctt.; misident., *nec* Sahlberg, 1819.
Gyrinus aeneus, sensu auctt.; misident., *nec* Stephens, 1829.

4.4-6.3 mm. Strongly resembling the two following species. Body oval, widest point at the middle; highest point usually in front of the middle (Fig. 88). Dorsal side black, rarely reddish, with a bronze sheen laterally. Ventral side metallic black; hypomera, epipleura, mesosternum and the fused gonocoxosternites rarely brownish. Legs yellow or reddish, femora and distal segments of mid and hind tarsi often darkened, claws of mid and hind tarsi black distally. Head, pronotum and elytra with rather strong micropunctuation, especially in females. Micro-reticulation hardly visible on head and pronotum. Inner rows of elytral punctures somewhat weaker than outer rows, especially anteriorly; intervals between the rows slightly elevated posteriorly; apical ellipsoid of punctures well marked; lateral impression of elytra narrow, without any dilation in front of the distinct outer apical angle (Fig. 51); females usually with elytra strongly micro-reticulated (discernable at × 40), especially towards apex; meshes polygonal, often slightly transverse or oblique (Fig. 27); males usually with micro-reticulation strongly obliterated, meshes more narrow, transverse or oblique (Fig. 26); specimens with greatly distorted micro-sculpture not uncommon. Outer distal angle of front tibia somewhat protruding (Fig. 92). Mesosternum with a narrow groove in the posterior half, sometimes weakly sulcate anteriorly (Fig. 47). Tergite VIII almost semicircular behind (Fig. 90). Male: penis narrow and parallel-sided distally for about one third of its length, with a rounded or truncate apex (Fig. 64). Female: inner part of basal edge of gonocoxae with a well delimited indentation (Fig. 77); basal muscular insertion area of gonocoxa of a characteristic shape.

Distribution. Rare in the eastern parts of Denmark (after 1950 only in B), more common in the west; not recorded from LFM. – Widespread and common in most parts of Sweden; not recorded from Gtl. and G. Sand. – Known from most Norwegian districts, north to TRi and Fø. – Almost all of Finland and adjacent parts of the USSR; generally common. – In Europe north to Britain, Fennoscandia and the USSR, south to northern France, German F.R. and D.R., and Poland; northern parts of the USSR, Siberia, Mongolia; *not* in Iceland, Greenland or N. America as stated by some authors.

Biology. Especially in oligotrophic lakes and at the edge of larger streams; rarely met with elsewhere (Svensson, 1969). In northern Fennoscandia only below the timber line (Huldén, 1983a). Adults often occur in large aggregations (schools), and local populations may number up to 10.000 individuals (Huldén, l.c.). The larva, not yet described, probably occurs in the summer. The adult hibernates. This species is the chief host of the ectoparasitic fungus *Laboulbenia fennica* Huldén (Huldén, 1983b).

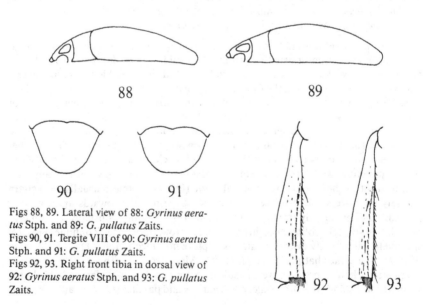

88   89

90   91

Figs 88, 89. Lateral view of 88: *Gyrinus aeratus* Stph. and 89: *G. pullatus* Zaits.
Figs 90, 91. Tergite VIII of 90: *Gyrinus aeratus* Stph. and 91: *G. pullatus* Zaits.
Figs 92, 93. Right front tibia in dorsal view of 92: *Gyrinus aeratus* Stph. and 93: *G. pullatus* Zaits.

92   93

4. *Gyrinus (Gyrinus) pullatus* Zaitsev, 1908
    Figs 28, 29, 52, 65, 78, 89, 91, 93.

Gyrinus (Gyrinus) pullatus Zaitsev, 1908, Russk. ent. Obozr. 7: 244.

4.5-6.0 mm. Strongly resembling the preceding and the following species, from which it differs in the following characters: body widest in the middle; highest point of body usually also in the middle (Fig. 89). Inner rows of elytral punctures usually not so weak anteriorly; intervals between the inner rows of punctures almost even; lateral impression narrow, at most with a very slight dilation in front of the poorly marked outer apical angle (Fig. 52); males usually with the micro-reticulation more reduced than in males of *aeratus* (Fig. 28). Outer distal angle of front tibia hardly protruding (Fig. 93). Tergite VIII more broadly rounded behind than in *aeratus* (Fig. 91). Male: the narrow distal part of penis only about one fourth the length of penis (Fig. 65); penis rounded apically. Female: inner part of basal edge of gonocoxae with a well delimited indentation (Fig. 78); basal muscular insertion area of gonocoxa of a characteristic shape.

Distribution. Not in Denmark or Norway. – Sweden: only in Nb: Pirriläjärvi NNW of Karungi, 15.vi.1941 (Brinck) (Svensson, 1982) and Blåmisusjön near Råneå, 1983 (Rydgård *et al.,* 1985). – Finland: mainly in the southeast, recorded from: Ab, N, Ka, St, Ta, Sa, Tb, Sb, Kb and Ok. – Adjacent parts of the USSR: Vib, Kr. – East Siberia, but probably more widely distributed in the USSR, Manchuria, Korea.

Biology. In dystrophic lakes and running water, according to Huldén (1983a); Rutanen (1976) has taken it in small woodland pools, often associated with *minutus* and *aeratus.* Adults sometimes occur in large local populations, up to 5.000 individuals (Huldén, 1983a). Not much is known about its life-cycle, and the larva has not yet been described.

5. *Gyrinus (Gyrinus) marinus* Gyllenhal, 1808
    Figs 12, 30-32, 37, 53, 66, 79.

Gyrinus dorsalis Gyllenhal, 1808, Ins. Suec. 1: 142.
Gyrinus marinus Gyllenhal, 1808, Ins. Suec. 1: 143.
Gyrinus aeneus Stephens, 1829, Ill. Brit. Ent., Mandib. 2: 95.
Gyrinus anthracinus Sturm, 1836, Deutschl. Faun. 5, Ins. 10: 102.
Gyrinus lembus Schiødte, 1841, Gen. Spec. Danm. Eleuth. 1: 563.
Gyrinus marinus var. *epipleuralis* Munster, 1925, Norsk ent. Tidsskr. 2: 32.

4.5-8.0 mm. Strongly resembling the two preceding species, from which it differs in the following characters: body usually widest behind the middle, and with the highest point near the middle. Micro-reticulation usually more reduced (particularly in southern specimens), but very strongly sculptured specimens do occur. Intervals between the inner elytral rows of punctures slightly or not elevated posteriorly; lateral impression usually rather wide, with a dilation before tapering towards the outer apical angle (Fig. 53); outer apical angle rounded, completely effaced; meshes of the elytral micro-reticulation usually narrow and oblique (Figs 30, 31), rarely more isodiametric. Outer distal angle of front tibia not or only slightly protruding. Tergite VIII rather broadly rounded behind, as in *pullatus*. Male: sides of penis tapering almost throughout its length; apex of penis pointed or very narrowly rounded (Fig. 66). Female: basal edge of gonocoxae with hardly any indentations (Fig. 79).

Distribution. Common all over Denmark. – Sweden: recorded from nearly all districts, more common in the south. – Norway: along the southern coast, west to Ry; in the north known from Fi and Fø. – Widespread in Finland and the adjacent part of the USSR, but not recorded from a few northern districts. – Most of Europe, north to Britain, Fennoscandia and the USSR, south to northern Spain, Italy and the Balkans; northern and central parts of the USSR, the Caucasus, Turkestan, Siberia, Mongolia, Kamtchatka; *not* in Greenland or N. America as stated by some authors.

Biology. Chiefly found in stagnant water, though occasionally met with in the backwaters of streams. It seems to prefer ponds and lakes with open, calm areas of surface, and generally avoids habitats with more dense vegetation at the surface, such as marshes. In the north it does not reach the timber line (Huldén, 1983a). Adults often occur in very large aggregations of individuals. The larva has been taken in late spring and summer, and was first described by Schiødte (1862). Freshly emerged young adults have been observed in July by West (1942). The adult hibernates clinging to submerged vegetation. The species has been observed to fly.

6. *Gyrinus (Gyrinus) colymbus* Erichson, 1837
Figs 33, 54, 67, 80.

Gyrinus colymbus Erichson, 1837, Käf. Mark Brandenb. 1: 191.
Gyrinus distinctus var. *strigulosus* Régimbart, 1891, Annls Soc. ent. Fr. 60: 677.
Gyrinus striolatus, sensu auctt.; misident., *nec* Fowler, 1887.

5.0-7.0 mm. Body oval, often fairly elongate, though always less than twice as long as wide. Dorsal side black with a bronze sheen laterally. Ventral side black, with hypomera, epipleura, mesosternum, the fused gonocoxosternites and sometimes other sclerites reddish. Legs, including claws, yellowish. Head, pronotum and elytra covered by short, deep, oblique strioles (Fig. 33). Micro-reticulation only discernable at high magnification, × 150. Inner rows of elytral punctures somewhat weaker than outer rows; apical ellipsoid of punctures well marked; lateral impression fairly narrow, hardly dilated before tapering towards the distinct outer apical angle (Fig. 54); meshes of elytral micro-reticulation narrow, oblique. Mesosternum with a narrow groove only in the posterior half. Male: penis somewhat dilated apically, with apex forming a blunt angle (Fig. 67). Female: gonocoxa short and wide with rounded sides (Fig. 80); basal edge of gonocoxa without indentations.

Distribution. Denmark: only one record from SZ (see note): Knudshoved, 1 ♂, 12.v.1910 (Hartvig Jensen) (Holmen, 1979). – Not in the rest of Fennoscandia. – France (including Corsica), German F.R. and D.R., Poland, Switzerland, Austria, Hungary,

Italy, the Balkans, Turkey, southern parts of European USSR; mediaeval subfossil remains have been discovered in England (Girling, 1983).

Biology. Not much is known about this species which seems rare everywhere, though most frequently met with in southeastern Europe. Guignot (1947) mentioned it from both fresh and saline water, and Franciscolo (1979) has taken it in swamps in relatively cold parts of Toscana. The Danish specimen was found in an area with sun-exposed ponds and marshes, very near the sea; this locality lies in an area which has comparatively hot and dry summers, and is known for its fauna of rare southern species.

Note. Three additional specimens have been found in a collection of Danish Gyrinidae, but no locality is stated on the label.

7. *Gyrinus (Gyrinus) distinctus* Aubé, 1836
Figs 13, 34, 55, 68, 81.

Gyrinus distinctus Aubé, 1836, *in* Dejean: Icon. Hist. Nat. Col. Eur. 5: 383.
Gyrinus caspius, sensu auctt.; misident., *nec* Ménétriés, 1832.
Gyrinus colymbus, sensu auctt.; misident., *nec* Erichson, 1837.

4.5-7.0 mm. Oval, fairly elongate, body usually about 1.9 times as long as wide. Dorsal side black with a bronze sheen laterally. Ventral side usually black, with hypomera, epipleura, mesosternum and the fused gonocoxosternites reddish; specimens from the southern and southeastern parts of the distributional area often have the ventral side completely yellowish or reddish. Legs, including claws, yellowish. Dorsal surface with a rather strong micro-punctuation, discernable at × 40 (Fig. 34). Micro-reticulation only discernable at high magnification, × 150. Inner rows of elytral punctures somewhat weaker than outer rows; apical ellipsoid of punctures well marked; lateral impression rather narrow, hardly dilated before tapering towards the distinct outer apical angle (Fig. 55); meshes of elytral micro-reticulation narrow, oblique; micro-punctures round. Mesosternum with a narrow groove only in the posterior half. Male: penis very wide, spoon-shaped (Fig. 68). Female: gonocoxa short and wide, with the outer edge rounded (Fig. 81); basal edge of gonocoxa at most with a weak indentation of the inner part.

Distribution. Widely distributed in Denmark, recorded from most districts. – In the southern part of Sweden, recorded from most districts north to Hls. – Norway: only in Ry: Jæren (Fritz Jensen). – Finland: only in the southern parts, recorded from all districts north to Oa. – Not in the adjacent parts of the USSR, but recorded from the Leningrad region. – Most of Europe, north to Britain, Fennoscandia and central parts of the USSR; the Azores, Cyprus, Sudan, Egypt, Israel, Lebanon, Syria, Turkey, Iran, Iraq, Transcaucasia, Afghanistan, Kashmir, Mongolia, China.

Biology. Found mainly at the edges of lakes and slowly flowing streams, especially in habitats with rather clear, base-rich water. Also known from brackish water. Adults usually occur in calm, shaded areas of the surface. Very little is known about its life-cycle, and the larva has not yet been described.

51

8. Gyrinus (Gyrinus) suffriani Scriba, 1855
Figs 56, 69, 82.

Gyrinus Suffriani Scriba, 1855, Stettin. ent. Ztg 16: 280.

4.0-6.5 mm. Rather similar to the two following species. Oval, fairly elongate. Highest point of body close to the middle. Dorsal side black with a bronze sheen laterally. Ventral side usually black, often with the hypomera, epipleura, mesosternum and the fused gonocoxosternites brownish; ventral side rarely reddish (S.European specimens). Legs, including claws, yellowish or reddish. Micro-reticulation and micropunctuation of dorsal surface only discernable at high magnification, ×150. Lateral border of pronotum distinct for almost its entire length, with the widest point (lateral view) closely behind the head. Rows of elytral punctures very strong anteriorly, especially towards the sides; apical punctures, including the ellipsoid, very weak, almost obsolete (Fig. 56); lateral impression rather narrow, slightly dilated and then narrowed before reaching the outer apical angle which is indistinct, somewhat effaced; meshes of elytral micro-reticulation narrow, oblique; micro-punctures round. Mesosternum only with a narrow groove posteriorly. Male: penis with a rather narrow distal part and a truncate or broadly rounded apex (Fig. 69). Female: gonocoxa rather long and narrow; inner part of basal edge of gonocoxae with a well delimited, rather wide indentation (Fig. 82).

Distribution. Rare and local in Denmark, not known from NWJ, NEJ, F and SZ. – Sweden: southern and central parts; sporadically in most districts north to Hls. – Norway: only in AAy: Risør, 1980 (Hagenlund) (Hagenlund, 1984). – Very rare in Finland, known only from Ab, N, Ka, Ta and Sb. – Not in adjacent parts of the USSR. – Most of Europe, north to Scotland and Fennoscandia; southern parts of European USSR; *not* in Eire, Spain or Portugal. Also records from Turkey, Syria, the Caucasus, and Transcaucasia. Seems to be rather rare everywhere.

Biology. Mainly found in oligotrophic or dystrophic bodies of water, both stagnant and running. Adults seem to prefer calm parts of the surface, close to the edge of the water or near tufts of vegetation. However, they probably spend much time below the surface. Oviposition has been observed in late spring, and under laboratory conditions development into adult is completed in about 56 days (Saxod, 1965a). Bertrand (1951) and Saxod (1965b) have described the larva, Saxod (1965b) the pupa.

9. Gyrinus (Gyrinus) natator (Linnaeus, 1758)
Figs 57, 70, 83, 94, 96; pl. 1: 7.

Dytiscus natator Linnaeus, 1758, Syst. Nat., ed. 10: 412.
Gyrinus mergus Ahrens, 1812, Neue Schr. naturf. Ges. Halle 2: 43.
Gyrinus marginatus Germar, 1824, Ins. Spec. Nov. 1: 32.
Gyrinus Wanckowiczi Régimbart, 1883, Ann. Soc. ent. Fr. 3: 157.
Gyrinus suffriani, sensu auctt.; misident., *nec* Scriba, 1855.

4.5-6.1 mm. Resembling the preceding and especially the following species, but differing in the following combination of characters: oval, highest point of the body near the middle (Fig. 94). Ventral side black, often with the hypomera, epipleura, mesosternum and the fused gonocoxosternites brown (rarely reddish). Lateral border of pronotum, at least in Fennoscandian specimens, poorly delimited in a short section in front of the middle, just in front of the widest point of the border (lateral view) (Fig. 96). Inner rows of elytral punctures strong, only a little weaker than the outer rows (Fig. 57); apical ellipsoid of punctures well marked. Male: penis only narrowed a little distally, apex truncate, weakly incised in the middle (Fig. 70); the sperm-groove on the ventral side of penis without a constriction about a third of the way from the tip of the groove to the base of penis. Female: inner part of basal edge of gonocoxae without a wide, well delimited indentation, but with a rather large projection or denticle on each gonocoxa (Fig. 83).

Distribution. Denmark: not in Jutland and NWZ, rare in the remaining districts. – Rather common in the south of Sweden, more scattered in the north. – Norway: only a few records from the southeast, Ø and AK. – Finland: in most districts south of the arctic circle, but not common. – Adjacent parts of the USSR: Vib, Kr. – Eire, England, France, Belgium, the Netherlands, German F.R. and D.R., Poland, Czechoslovakia, Hungary, central and northern parts of European USSR; Siberia, China. Very rare in most parts of western Europe.

Biology. Chiefly found in smaller, slow-flowing streams, and in sun-exposed marshes and ponds. Adults seem to prefer calm areas of the surface without too much penetrating vegetation. The larva, not yet described, probably occurs in the summer. The adults hibernate, clinging to submerged vegetation. Flying specimens have been observed.

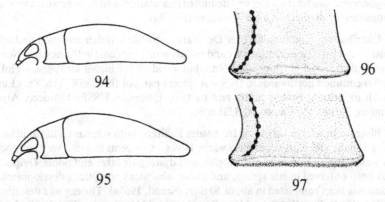

Figs 94, 95. Lateral view of 94: *Gyrinus natator* (L.) and 95: *G. substriatus* Stph.
Figs 96, 97. Pronotum in left lateral view of 96: *Gyrinus natator* (L.) and 97: *G. substriatus* Stph.

10. *Gyrinus (Gyrinus) substriatus* Stephens, 1829
   Figs 19, 22, 23, 35, 41, 42, 58, 71, 84, 95, 97.

Gyrinus substriatus Stephens, 1829, Ill. Brit. Ent., Mandib. 2: 97.
Gyrinus cercurus Schiødte, 1841, Gen. Spec. Danm. Eleuth. 1: 564; see note.
Gyrinus natator var. *corpulentus* Schatzmayr, 1903, Wien. ent. Ztg 22: 172; preocc.,
   nec Régimbart, 1883.
Gyrinus natator var. *oblitus* Sharp, 1914, Entomologist's mon. Mag. 50: 133.
Gyrinus natator substriatus a. *Schatzmayri* Ochs, 1927, Koleopt. Rdsch. 13: 36; re-
   placement name for *corpulentus* Schatzmayr, 1903.
Gyrinus natator var. *fowleri* Omer-Cooper, 1930, Entomologist's mon. Mag. 66: 74.
Gyrinus natator, sensu auctt.; misident., *nec* (Linnaeus, 1758).
Gyrinus bicolor, sensu auctt.; ? misident., ? *nec* Fabricius, 1787; see note under *G.
   paykulli.*
Gyrinus mergus, sensu auctt.; misident., *nec* Ahrens, 1812.

5.0-7.0 mm. Resembling the two preceding species, especially *natator,* from which it differs in the following combination of characters: oval, highest point of body in front of the middle, just behind pronotum (Fig. 95). Dorsal colour and micro-sculpture usually as in the two preceeding species, but reddish specimens and specimens with larger, often irregular micro-punctures are known to occur. Ventral side usually largely black, with the hypomera, epipleura, mesosternum and the fused gonocoxosternites reddish (rarely brownish or black). Lateral border of pronotum distinct for almost its entire length, with the widest point (lateral view) closely behind the head (Fig. 97). Inner rows of elytral punctures usually very weak, in the middle sometimes almost obsolete; outer rows stronger (Fig. 58); apical ellipsoid of punctures well marked. Male: penis only narrowed a little distally, apex truncate, weakly incised in the middle (Fig. 71); the sperm-groove on the ventral side of penis distinctly constricted about a third of the way from the tip of the groove to the base of penis. Female: inner part of basal edge of gonocoxae without a wide, well delimited indentation, and at most with a very small projection or denticle on each gonocoxa (Fig. 84).

Distribution. Common all over Denmark. – Sweden: rather common and widely distributed in the south, more rare and mainly along the coast in the north. – Norway: widely distributed northwards to Nnø, but avoiding the higher altitudes. – Finland: rather common northwards to ObS. – Adjacent parts of the USSR: Vib, Kr. – Europe north to Britain, Fennoscandia and parts of European USSR; Morocco, Algeria, Tunisia, Turkey, the Caucasus, S.Siberia.

Biology. In a large variety of freshwater habitats, both stagnant and running, and occasionally met with in brackish water. However, it seems to avoid very large bodies of water, and to be rare in oligotrophic and dystrophic lakes and pools. Oviposition has been observed in late spring, and under laboratory conditions development into adult has been completed in about 50 days (Saxod, 1965a). The egg was described by Saxod (1964); Bott (1928), and later Saxod (1964, 1965), described the larva; the pupa was described by Saxod (1965b). This species has often been observed to fly.

Note on synonymy. *Gyrinus cercurus* Schiødte, 1841 was probably described on specimens of both *G. natator* (L.) and *G. substriatus* Stph., and a specimen of each species labelled *"cercurus"* by Schiødte are present in the Zoological Museum, University of Copenhagen. I have followed the general view, that *G. cercurus* Schiødte, 1841 is a junior synonym for *G. substriatus* Stephens, 1829, as no lectotype has been designated.

11. *Gyrinus (Gyrinus) caspius* Ménétriés, 1832
Figs 59, 72, 85.

Gyrinus Caspius Ménétriés, 1832, Cat. Rais. Cauc.: 142.
Gyrinus elongatus Aubé, 1836, *in* Dejean: Icon. Hist. Nat. Col. Eur. 5: 384; preocc., ? *nec* Marsham, 1802.
Gyrinus angustatus Aubé, 1836, *in* Dejean: Icon. Hist. Nat. Col. Eur. 5: 387.
Gyrinus celox Schiødte, 1841, Gen. Spec. Danm. Eleuth. 1: 565.
Gyrinus bicolor, sensu auctt.; ? misident., ? *nec* Fabricius, 1787; see note under *G. paykulli*.
Gyrinus mergus, sensu auctt.; misident., *nec* Ahrens, 1812.
Gyrinus distinctus, sensu auctt.; misident., *nec* Aubé, 1836.

4.9-7.5 mm. Resembling the following species. Very elongate, body more than twice as long as wide, sides often parallel in the middle. Dorsal side black with a bronze sheen laterally. Ventral side largely black, with the hypomera, epipleura, mesosternum and the fused gonocoxosternites reddish. Legs, including claws, yellowish. Dorsal microsculpture only discernable at high magnification, × 150. Inner rows of elytral punctures somewhat weaker than outer rows; apical ellipsoid of punctures well marked; lateral impression rather narrow, only weakly dilated and then narrowed before reaching the very distinct outer apical angle (Fig. 59); meshes of elytral micro-reticulation narrow, oblique; micro-punctures round. Mesosternum with only a narrow groove posteriorly. Male: penis slightly dilated distally, with a broadly rounded apex (Fig. 72). Female: gonocoxae rather short and wide, distal edges rounded (Fig. 85).

Distribution. Denmark: all disticts, mainly found near the coast. - Sweden: recorded from Sk., Hall., Sm. and Upl., mainly near the coast. - Norway: known only from a few records in AAy. - Not in East Fennoscandia, where previous records apply to *G. paykulli* (Huldén, 1983a). - Most of Europe, north to Britain and Fennoscandia; in the south of European USSR; Morocco, Algeria, Turkey, Iraq, the Caucasus, eastwards to Tibet and China (Sinkiang).

Biology. In northern Europe mainly found near the sea, often in brackish water. Inland records are more scarce, chiefly from running water. Adults are usually taken in beds of *Phragmites* or *Scirpus*, even where this vegetation is quite dense, but also in more open areas of the surface. Oviposition has been observed in May and July. Bertrand (1951) provides a short description of the young larva. The adult hibernates clinging to submerged vegetation. The species is known to fly (Jackson, 1973).

12. *Gyrinus (Gyrinus) paykulli* Ochs, 1927
Figs 60, 73, 86.

Gyrinus Paykulli Ochs, 1927, Koleopt. Rdsch. 13: 39.
Gyrinus paykulli a. *ochsi* Franck, 1932, Ent. Bl. Biol. Syst. Käfer 28: 133.
Gyrinus bicolor, sensu auctt.; ? misident., ? *nec* Fabricius, 1787; see note.
Gyrinus celox, sensu auctt.; misident., *nec* Schiødte, 1841.

5.5-8.5 mm, - our largest gyrinid. Resembling *caspius,* especially in the narrow shape
of the body, but differing as follows: ventral side largely black, with the hypomera,
epipleura, mesosternum and the fused gonocoxosternites brownish or black, rarely
reddish. Outer apical angle of elytra rounded, almost completely effaced (Fig. 60).
Male: penis very narrow distally, with a pointed or narrowly rounded apex (Fig. 73).
Female: gonocoxa usually rather long and narrow, with the distal edge almost obliquely truncate (Fig. 86); specimens with gonocoxae resembling those of *caspius,* are
however, said to occur.

Distribution. All of Denmark, though not particularly common. - Mainly in the
southern and central parts of Sweden, with scattered records as far north as Vb. - Only
in the southwest and southeast of Norway, recorded from VAy, Ry, HOy and HEs. -
Fairly common in southern Finland, northwards to Om and Ok. - Adjacent parts of
the USSR: Vib and Kr. - In most of Europe, north to Britain, Fennoscandia and parts
of European USSR, south to northern Spain, Italy and Greece; Turkey, the Caucasus,
eastwards to eastern Siberia, Mongolia and China (Sinkiang).

Biology. Mostly in pools and dystrophic lakes, more rare in brackish water and
streams than the preceeding species. Adults mainly found in, or near, beds of *Phragmites* or *Scirpus,* but also in more open areas of the surface. Oviposition has been observed in late spring, and under laboratory conditions development into adult has
been completed in about 56 days (Saxod, 1965a). The larva was first described by Bertrand (1951); Saxod (1965b) gives a description of the larva and the pupa. The adult
hibernates clinging to submerged vegetation.

Notes on synonymy. Fabricius (1787) originally described *G. bicolor* on specimen(s)
of same size as *natator,* with a reddish ventral side, and collected in Sweden by Leske.
The type material has not since been discovered, and the identity of the species remains
uncertain. The only Fennoscandian species which normally has a reddish ventral side
is the small *G. minutus* described by Fabricius (1798). Paykull (1798) interpreted
Fabricius' description as the species now known as *G. paykulli* Ochs. A few years later
Fabricius (1801) added to the description that the species is elongate, but this character
may have been taken from Paykull's description. Other authors have considered
Fabricius' species to be identical with *G. natator* (L.), *G. substriatus* Stph. or *G.
caspius* Mén. All these may, like *paykulli,* have parts of the ventral side reddish. Any-
way, both *paykulli* and *natator* are present in Fabricius' collection, but under the name
"*natator*". There is also the possibility that the original specimens of *bicolor* were
wrongly interpreted as Swedish, as Leske also worked on many Italian insects. If this

is the case, the specimens might well be conspecific with *G. urinator* Ill. This is the European species which normally best fits the original description of *bicolor,* and it is very common in the Mediterranean area.

Olivier (1795), Paykull (1798) and Latreille (1807) never originally described species by the name *Gyrinus bicolor,* as stated by many authors; each of these authors merely interpreted Fabricius' description as one of the above mentioned species.

Gyrinus (Gyrinus) urinator Illiger, 1807
Figs 61, 74, 87.

Gyrinus Urinator Illiger, 1807, Mag. Insektenkunde 6: 299.
Gyrinus lineatus Stephens, 1829, Ill. Brit. Ent., Mandib. 2: 97.
Gyrinus bicolor, sensu auctt.; ? misident., ? *nec* Fabricius, 1787; see note under *G. paykulli.*

4.8-7.8 mm. Rather broadly oval. Dorsal side black (rarely reddish) with a bronze sheen laterally, and with bronze stripes following the elytral rows of punctures. Ventral side and legs reddish or yellowish, sometimes darkened; claws yellowish or reddish. Dorsal micro-sculpture only discernable at high magnification, × 150. Inner rows of elytral punctures almost obsolete in the middle, much weaker than outer rows; apical ellipsoid of punctures well marked; lateral impression poorly delimited, very wide, dilated posteriorly and then tapering towards the distinct outer apical angle (Fig. 61); meshes of elytral micro-reticulation narrow, oblique; micro-punctures round. Mesosternum with a very fine median groove in the posterior half. Male: penis only a little narrowed distally, the broadly rounded apex with a median notch (Fig. 74). Female: gonocoxae rather short and wide; inner part of their basal edge with a large, well delimited indentation (Fig. 87).

Distribution. Not in Denmark or Fennoscandia. - A largely Mediterranean species, in Europe occurring northwards to Eire, England, France, German F.R. and D.R., Poland and parts of European USSR; ? Madeira, the Canaries, N.Africa (except Egypt), Turkey.

Biology. Mostly found in small streams, but, particularly in the south, also in various stagnant habitats, including ponds, rock-pools and brackish water. Adults usually found in small, open areas of the surface. In Britain most adult specimens have been caught submerged, but this is not usual further south (Balfour-Browne, 1950). In the Atlas Mountains it has been recorded from an altitude of 1200 m. Bertrand (1951) gives a description of the young larva.

Genus *Aulonogyrus* Motschulsky, 1853

Aulonogyrus Motschulsky, 1853, Hydrocanth. Russ.: 9; see note.
Type species: *Gyrinus strigipennis* Suffrian, 1842 [= *Aulonogyrus (Aulonogyrus) concinnus* (Klug, 1834)], by subsequent designation (Ochs, 1930); see note.

Small or medium-sized species. Body oval or elongate, often rather depressed. Dorsal side largely black, sometimes with a yellow lateral border on pronotum and elytra, and often with a bright metallic sheen of bronze, purple, blue and green. Our species have a short labrum with longitudinal furrows, and have the dorsal eyes anterior to the ventral eyes, reaching forwards as far as the antennal base. Antenna 9-segmented. Pronotum usually without distinct transverse grooves. Elytra with round or elongate punctures forming 11 rows and often an apical ellipsoid; the inner rows are often very weak, and the outermost row is placed in the lateral impression. The punctures follow shallow, often strongly micro-reticulated furrows which are often confluent laterally. The intervals between the furrows are shining, with more weakly impressed micro-reticulation and scattered fine punctures. The meshes of the elytral micro-reticulation are polygonal, discernable at low magnification, × 40. Tergite VII strongly trilobed behind (Fig. 100).

The genus contains about 55 species, the majority of which are found in Africa and Madagascar. Two are known from Europe; one of these penetrates eastwards into Mongolia and China. One species lives in southern India and Sri Lanka, and two in the Australian region. The genus has been revised by Brinck (1955), who divided it into 6 subgenera, of which *Aulonogyrus* s. str. occurs in Europe.

Species of *Aulonogyrus* are chiefly found in running water, but some may also occur in stagnant habitats.

Note on synonymy. Motschulsky (1853) created the genus *Aulonogyrus* for two species, viz., *striatus* Fabr. and *strigipennis* Suffrian, both sensu Motschulsky. He gave the Russian distribution for the two species, but otherwise no descriptions were provided, neither for the genus nor for the species. However, according to the rules of the International Commission on Zoological Nomenclature (art. 12, b, 5), the genus should certainly be attributed to Motschulsky (l.c.), as also stated by Ochs (1930), and not to Régimbart (1883), as stated by J. Balfour-Browne (1945), Brinck (1955) and others.

Motschulsky's specimens were later examined by Zaitsev (1915). The Russian specimens all belong to *A. (Aulonogyrus) concinnus* (Klug, 1834) [= *Gyrinus strigipennis* Suffrian, 1842], the only member of the genus known from Russia (Zaitsev, 1953). The true *A. (Aulonogyrus) striatus* (Fabricius, 1792) was present (and correctly identified .by Motschulsky) only among Mediterranean specimens that were not included in the original description of the genus.

Ochs (l.c.) designated *concinnus* Klug as the type species of *Aulonogyrus* Motschulsky, 1853, and synonymized the species with *strigipennis* Suffr. and *striatus* Motsch., 1853. As the name *concinnus* was not originally included in the genus, I consider *Gyrinus strigipennis* Suffrian, 1842 [= *A. (Aulonogyrus) concinnus* (Klug,

1834)] to be the type species of *Aulonogyrus* Motschulsky, 1853, by subsequent designation (Ochs, 1930).

Key to species of *Aulonogyrus*

1 Elytral interval 8, and usually also interval 10, depressed, normally without shining, weakly micro-reticulated patches in the anterior half (Fig. 98). Male: penis very wide distally (Fig. 101). Female: inner part of basal edge of gonocoxae with a large indentation (Fig. 103). Length: 5.1-7.0 mm ..... *concinnus* (Klug)

– Elytral intervals 8 and 10 elevated, often shining, weakly micro-reticulated, in the middle for most of their length (Fig. 92). Male: penis narrower distally (Fig. 102). Female: inner part of basal edge of gonocoxae with a rather small indentation (Fig. 104) Length: 5.5-8.0 mm .... *striatus* (Fabricius)

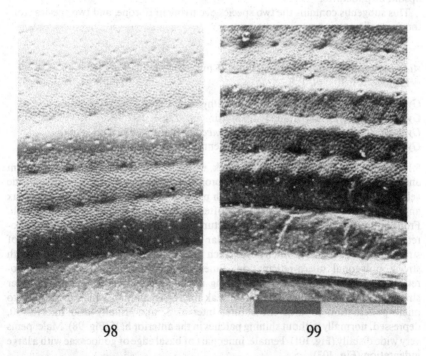

98 99

Figs 98, 99. Lateral part of left elytron of 98: *Aulonogyrus concinnus* (Klug) and 99: *A. striatus* (F.); electroscan micrographs.

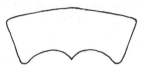

Fig. 100. Tergite VII of *Aulonogyrus striatus* (F.).

Subgenus *Aulonogyrus* s. str.

Aulonogyrus sg. *Aulonogyrus* s. str.

Head not particularly depressed. Dorsal eyes distinctly anterior to ventral eyes and rather widely separated from each other (by about 1½ times the width of one eye). The ridge bordering the top of the ventral eye not developed into a broad and flat epipleuroid disc. Labrum short, strongly transverse, with a fringe of long hairs inserted in a dorsal furrow behind the anterior edge, and with longitudinal furrows. Anterior edge of pronotum sinuate, but not lobiform in the middle. Elytra with an apical ellipsoid of punctures.

This subgenus contains the two species occurring in Europe, and two species from the Australian region.

Aulonogyrus (Aulonogyrus) concinnus (Klug, 1834)
Figs 98, 101, 103; pl. 1: 8.

Gyrinus concinnus Klug, 1834, Symbol. Physic. 4: t. 34.
Gyrinus strigipennis Suffrian, 1842, Stettin. ent. Ztg 3: 226.
Gyrinus striatus, sensu auctt.; misident., *nec* Fabricius, 1792.
Gyrinus abdominalis, sensu auctt.; misident., *nec* Aubé, 1838.

5.1-7.0 mm. Oval. Dorsal side largely black, with a yellow lateral border on pronotum and elytra, and with a metallic sheen of bronze, purple, blue and green. Ventral side yellowish to a varying extent, usually with parts of the head, meso- and metathorax and the abdominal sternites darkened (often almost black). Legs largely yellowish. Frons and vertex with fine irregular punctures and wrinkles, without distinct micro-reticulation. Pronotum with dense irregular punctuation, and with lateral patches of strong micro-reticulation which never reach the mid-line. Elytral lateral furrows with strong polygonal micro-reticulation; inner furrows much weaker, with micro-reticulation more obliterated; the shining interspaces between the furrows rather strongly and densely punctured, with weak micro-reticulation which dissolves into micro-punctuation towards the suture; interval 8, and usually also interval 10, depressed, normally without shining patches in the anterior half (Fig. 98). Male: penis very wide distally (Fig. 101). Female: inner part of basal edge of gonocoxae with a large indentation (Fig. 103).

Distribution. Not recorded from Denmark or Fennoscandia. – In Europe mainly

60

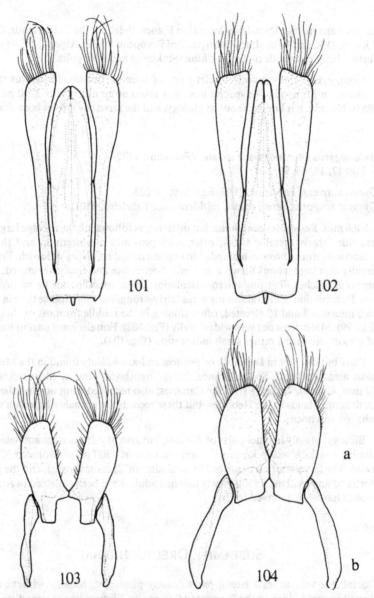

Figs 101, 102. Male genital sclerites (dorsal view) of 101: *Aulonogyrus concinnus* (Klug) and 102: *A. striatus* (F.).

Figs 103, 104. Female genital sclerites of 103: *Aulonogyrus concinnus* (Klug) and 104: *A. striatus* (F.); a: gonocoxae; b: tergal halves IX.

Mediterranean, distributed northwards to France, Belgium, the Netherlands, German F.R. and D.R. and Poland; southern parts of European USSR; Algeria, ? Egypt, Syria, Turkey, Iraq, eastwards to Tibet, China (Sinkiang) and Mongolia.

Biology. Especially found in quiet parts of streams, but also in various stagnant habitats. In Mongolia the species has been taken at an altitude of 3200 m. (Ochs, 1953). Not much is known about its biology, and the larva has not yet been described.

Aulonogyrus (Aulonogyrus) striatus (Fabricius, 1792)
Figs 17, 18, 45, 99, 100, 102, 104.

Gyrinus striatus Fabricius, 1792, Ent. Syst. 1: 203.
Gyrinus strigosus, sensu auctt., misident., *nec* Fabricius, 1801.

5.5-8.0 mm. Resembles *concinnus,* but differing as follows: on the average larger. Ventral side largely metallic black, often with pro- and mesosternum and the fused gonocoxosternites brownish or reddish; hypomera and epipleura yellowish. Frons and usually also impressions between eyes micro-reticulate and finely punctured. Lateral pronotal patches of strong micro-reticulation often extending to the pronotal midline. Punctuation of elytral shining areas fairly strong, but more sparse than in *concinnus;* intervals 8 and 10 elevated, often shining in the middle for most of their length (Fig. 99). Male: penis not very wide distally (Fig. 102). Female: inner part of basal edge of gonocoxae with a rather small indentation (Fig. 104).

Distribution. Not in Denmark or Fennoscandia. – Mainly found in the Mediterranean area: Portugal, Spain, France, Italy, Yugoslavia, Greece, Morocco, Algeria, Tunisia, Cyprus, Turkey, Syria, the Canaries; also recorded from northwestern Scotland: some islands in the Hebrides, but these records are probably based on wrongly labelled specimens.

Biology. Mainly in quiet parts of streams, but also in various stagnant habitats, including brackish water. Recorded from altitudes of 1200 m. in Morocco (Guignot, 1946). The larva was first described by Schiødte (1872) (as *strigosus*), and the pupa by Bertrand and Vaillant (1950). Newly hatched adults have been collected in April. The species has been observed to fly.

SUBFAMILY ORECTOCHILINAE

Dorsal side with at least lateral parts densely pubescent. Maxilla without a galea. Scutellum not hidden in the Fennoscandian genus. Elytra without very distinct rows of large punctures. Metasternum not divided by a transverse suture. Gonocoxosternites fused, very elongate in outline. Sternite VII and the fused gonocoxosternites with a longitudinal row of long hairs medially.

The subfamily comprises three large genera, viz., *Orectochilus* which is represented in Fennoscandia, *Gyretes* from the Neotropical and Nearctic regions, and *Orectogyrus* from the Afrotropical region.

Genus *Orectochilus* Dejean, 1833

Orectochilus Dejean, 1833, Cat. Coll. Coleopt. Dejean, ed. 2: 59, (see note).
   Type species: *Gyrinus villosus* Müller, 1776 [= *Orectochilus (Orectochilus) villosus* (Müller, 1776)], by subsequent designation (Hope, 1838).

Small, medium-sized or large species. Body oval or elongate. Dorsal side in our species strongly convex, black or brownish, without metallic sheen. Labrum rather long, semi-circular in many species, without longitudinal furrows. Dorsal eyes hardly anterior to ventral eyes. Antenna 8-11-segmented. Pronotum without distinct transverse grooves. Scutellum not hidden by pronotum or elytra. Elytra with rather evenly punctured parts, and with hairs originating from the punctures; punctures not arranged in distinct rows. Elytra with a polygonal micro-reticulation which is weak or partly obliterated in many species, usually discernable at about × 100. Micro-punctuation usually absent. Lateral parts of mid coxae gradually and distinctly attenuated towards the more or less narrow apex. Tergite VII not trilobed behind.
   The species are chiefly found in streams and along the edges of larger lakes.
   Orectochilus probably comprises close to 200 species (Franciscolo, 1979), but a revision of this large genus is not available. The majority of species belong to subg. *Patrus* Aubé which is largely Oriental, but with one representative in Zaire. The few remaining species belong to subg. *Orectochilus* s. str., represented in the Oriental and Palearctic regions.

Note on synonymy. Dejean (1833) was the first author to publish the name *Orectochilus* (attributing it to Eschscholtz). As he included in it available species names, he should be considered the first author of the genus, even in the absence of a description (ICZN rules: art. 12, b, 5). A number of other original authors have been proposed over the years, but their works are all published later than that of Dejean (l.c.).

Subgenus *Orectochilus* s. str.

Orectochilus sg. *Orectochilus* s. str.

Pronotum and elytra normally entirely pubescent, without glabrous spaces medially.
   Antenna of the Fennoscandian species 11-segmented.
   This small Oriental and Palearctic subgenus contains the only Fennoscandian species of the genus.

13. *Orectochilus (Orectochilus) villosus* (Müller, 1776)
Figs 36, 44, 105, 106; pl. 1: 9.

Gyrinus villosus Müller, 1776, Zool. Dan. Prodr.: 68.
Gyrinus Viola aquatica Modeer, 1776, Physiogr. Sälsk. Handl. 1: 160.
Gyrinus Modeeri Marsham, 1802, Ent. Brit. 1: 100.
Orectochilus Seidlitzi Jakobson, 1907, Zhuki Ross. Zapadn. Evr. 5: 439.

5.0-8.0 mm. Elongate, strongly convex, laterally compressed. Dorsal side black or brownish. Ventral side, including legs, yellowish, often darkened. Head, pronotum and elytra densely punctured and pubescent, with a weak polygonal micro-reticulation which is strongest towards the sides, discernable at about × 100 (Fig. 36); micro-punctuation not distinct. Male: penis gradually narrowed into a pointed apex (Fig. 105). Female: gonocoxae long and narrow; spermatheca well sclerotized, boat-shaped (Fig. 106).

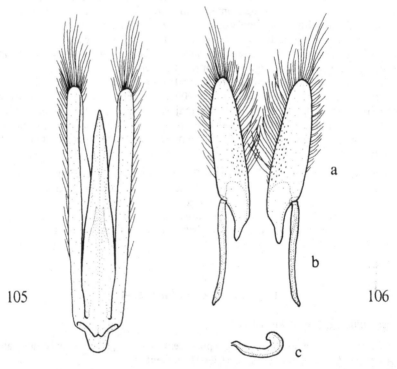

105

106

Figs 105, 106. Male genital sclerites (105) and female genital sclerites (106) of *Orectochilus villosus* (Müll.); a: gonocoxae; b: tergal halves IX; c: spermatheca.

64

Distribution. All of Denmark, excepting LFM. – Widely distributed in southern Sweden, north to Dlr. and Gstr. – Only in the southwest and southeast of Norway, recorded from Bø, Ve, VAy and Ry. – In the south of Finland, north to St, Tb and Sa. – Adjacent parts of the USSR· Vib. – Most of Europe, north to Britain and Fennoscandia; European USSR; Morocco, Algeria, Egypt, Israel, Syria, Turkey, eastwards to Siberia, China and Japan.

Biology. In streams, especially the larger ones, and in lakes. Occasionally also in brackish water along the coast. Seems to stand rather polluted water (Huldén, 1983a). The adult is mainly active during the night, and often hides under stones and the like during daytime. In Fennoscandia the adults mainly appear from June to September, and probably do not hibernate. Oviposition takes place during the summer, and the larva hibernates in the water; newly hatched adults appear in late spring and summer (Balfour-Browne, 1950). The eggs were described by Berg (1948), the larva by Schiødte (1864), and the pupa by Lesne (1902). The species has been recorded from altitudes up to 1600 m. in Morocco (Kocher, 1958). It is probably unable to fly due to skeletal and muscular reductions (Larsén, 1966).

FAMILY HALIPLIDAE
Morphology of the adult
(Figs 107-119).

Body adapted for swimming or crawling in water bodies. Body length varies from 1.5-5.0 mm. (2.0-4.6 mm. in the Fennoscandian species). Front and mid tarsi of males with one, two or three basal segments dilated and provided with tufts of sucker-hairs on the underside.

Head (Figs 107, 108).
Prognathous and generally fairly small. Eyes large and somewhat protruding, not divided. Labrum transverse and narrow; usually curved inward anteriorly with a marginal fringe of hairs; often punctured on its dorsal surface. Clypeus wider than labrum, transverse and usually curved inward anteriorly. Clypeus fixed posteriorly to frons without any distinct suture. Frons and vertex fused without any visible sutures; forming a fold above the antennal base. Ventral side of head, behind the mouthparts, composed of a gula medially and the genae laterally. Sides of head behind the eyes often with a number of vertical furrows of taxonomical importance (Ruhnau, in litt.).

Antennae (Figs 107, 108).

Moderately long, 11-segmented and filiform. Basal two segments slightly wider than the following segments; distal segments usually longer than proximal segments.

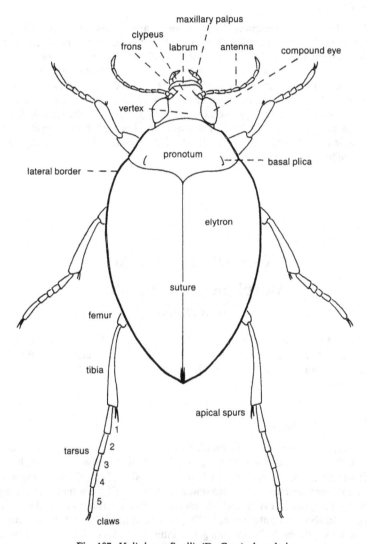

Fig. 107. *Haliplus ruficollis* (De Geer), dorsal view.

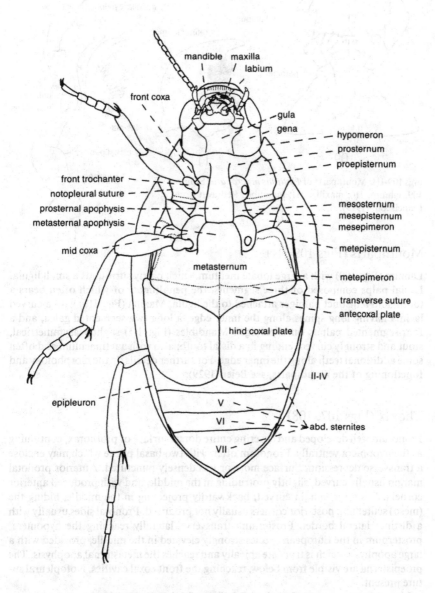

Fig. 108. *Haliplus flavicollis* Sturm, ventral view.

67

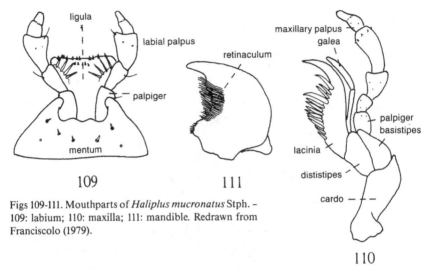

Figs 109-111. Mouthparts of *Haliplus mucronatus* Stph. –
109: labium; 110: maxilla; 111: mandible. Redrawn from
Franciscolo (1979).

Mouthparts (Figs 109-111).

Labium (Fig. 109) with a large lobate mentum which partly surrounds a small ligula.
Labial palps composed of three segments, the penultimate of which often bears a
tooth, and a distinct palpiger attached to the ligula. Maxilla (Fig. 110) with a curved
lacinia having long spines along the inner edge, a long two-segmented galea, and a
four-segmented palpus on a palpiger. Mandibles (Fig. 111) slightly asymmetrical,
stout and strongly curved, ending in a distal tooth; also with a retinaculum and often
some additional teeth along the inner edge. For further details on the morphology and
functioning of the mouthparts, see Beier (1929).

Thorax (Figs 107, 108)

Pronotum well developed and covering entire dorsal surface of prothorax, continuing
as the hypomera ventrally. Pronotum often with two basal plicae which may enclose
a transverse depression. Surface more or less densely punctured. Anterior pronotal
margin usually curved, slightly protruding in the middle, and with produced anterior
corners. Posterior margin curved, backwardly projecting in the middle, hiding the
(meso) scutellum; posterior corners usually not produced. Pronotal sides usually with
a distinct lateral border. Prosternum transverse, laterally reaching the hypomera;
prosternum in the European species strongly elevated in the middle, provided with a
large apophysis which is truncate apically and reaches the metasternal apophysis. The
proepisterna are visible from below, reaching the front coxal cavities. Notopleural suture present.

Mesothorax small. Its dorsal surface is completely covered by the elytra and the pronotum. Mesosternum small, partly covered by the prosternal apophysis. Both mesepimera and mesepisterna visible from below.

Metathorax large, dorsally covered by the elytra. Most of the ventral side is occupied by the large metasternum and the hind coxae of which the latter reach the epipleura.

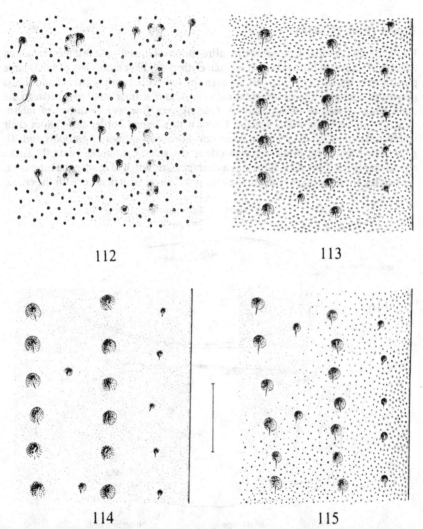

Figs 112-115. Dorsal elytral surface of *Haliplus* spp. - 112: *H. confinis* Stph.; 113: *H. apicalis* Thoms.; 114: *H. ruficollis* (De Geer) ♂; and 115: same species ♀. Scale 0.1 mm. Michael Hansen del., drawings based on electroscan micrographs.

Metasternum transverse, transversely divided by a suture separating the antecoxal plate from the anterior part of the sclerite. Metasternum separated laterally from the epipleura by the metepisterna and metepimera; median part produced anteriorly as the metasternal apophysis.

Elytra (Figs 112-116)

Free and well sclerotized, covering the entire abdomen dorsally. At rest meeting along the elytral suture. In European species each elytron has 10 or more rows of fairly large punctures (Figs 112-115). Intervals between the rows often with scattered additional punctures, and usually with irregular rows of somewhat finer punctures between the main rows as well as presuturally. A fine, or very fine, evenly dispersed micro-punctuation occurs in many species. Elytra with a lateral border throughout their length (serrate in *Brychius* and in some non-European species of *Haliplus* and *Algophilus*). Elytra bent down- and inwards at the sides, continuing ventrally as the elytral epipleura. Each epipleuron has a characteristic notch to receive the hind coxa, connecting the air-storage beneath the hind coxal plates with that beneath the elytra.

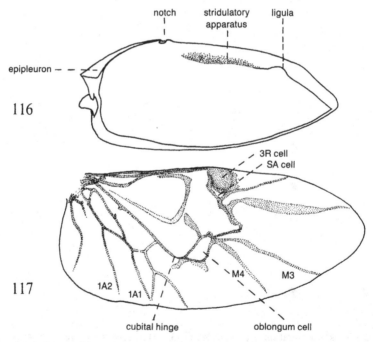

Fig. 116. Left elytron of *Haliplus wehnckei* Gerh. viewed from below. Redrawn from Beier (1929).
Fig. 117. Hind wing of *Haliplus fulvus* (F.).

Ventral side of each elytron with a lateral "ligula"; this extension fits into a depression of the dorsal side of the abdomen, fixing the elytron when the insect is not flying. Elytra without ventral patches of spicules for binding the hindwings. A few species are known to produce sounds by stridulation; in *Haliplus wehnckei* the stridulatory apparatus consists of a band of fine teeth near the lateral elytral edge (Fig. 116) and the edge of laterosternites II-IV (Beier, 1929).

Hindwings (Fig. 117)

Membranous, hyaline and well developed in the Fennoscandian members of the family (absent in the non-European genera *Algophilus* and *Apteraliplus*). Venation (nomenclature following Ward (1979)) is of an Adephagan type in which the M4 originates anterior to the middle of the Oblongum Cell (Fig. 117). A few other characteristics of the haliplid wing should be mentioned: Oblongum Cell broadly oval or in shape of a parallelogram; M4 originates close to M3; the 3R Cell larger than the SA Cell; a wedge-shaped Cell between 1A1 and 1A2 absent; subcubital binding patch of spicules absent. Wing-folding also of an Adephagan type, with the cubital hinge not very far removed from the Oblongum Cell; the folding itself takes place by the aid of tergal binding patches.

Legs (Figs 107, 108)

These are long and slender. Front and mid legs rather unmodified; front and mid coxae subglobular, movable; trochanter short, somewhat globular; femur widened in the middle; tibia and tarsus slender, subcylindrical; tarsus with five segments of which the distal one bears a pair of claws; these may be finely serrate or pectinate ventrally. Hind coxae very large and immovably fixed to ventral thoracic sclerites. Each coxa forms a large backwardly projecting plate which covers trochanter, basal part of femur, and ventral parts of the abdomen. The space enclosed by the coxal plates serves as an air storage, and is connected with the space enclosed by the elytra through the anterolateral corners of the coxae. Hind trochanter small, subtriangular; hind femur slender, dilated at both ends; hind tibia and tarsus long and slender, subcylindrical; tibiae with two apical spurs which are pectinate in many species, and tarsus with 5 segments of which the distal one bears a pair of claws. Legs, especially tibiae and tarsi, provided with rows of spines and swimming-hairs. The row of swimming-hairs which originates from a row of closely set punctures, the setiferous striole, on the upper face of hind tibia, is of diagnostic importance.

Abdomen (Figs 107, 108, 118)

Abdomen (Fig. 118) dorsally with 8 visible, partly membranous, tergites (segments I-VIII), and with a number of more or less sclerotized pleurites. These are totally covered by the elytra. Ventrally 6 sternites (II-VII) are exposed; sternite VII sometimes with a median furrow, pointed apically. Sternites II-VI fused without any visible sutures, and furthermore immovably fixed to hind coxae along their anterior margin. Lateral parts (laterosternites) of sternites II-V covered by the elytra. Tergite IX, gonocoxosternites, and sternites VIII-IX invaginated, more or less reduced, or modified to function as parts of the reproductive system. Pleurites I-VII and tergal margin VIII with a pair of spiracles on each segment.

Surface sculpture (Figs 112-115)

Body without visible micro-reticulation, but usually with very prominent punctuation which offers good diagnostic characters (Figs 112-115). Punctures of various size categories are often present in the same specimen. Larger punctures, discernable at about × 10, may be more or less evenly dispersed on some parts of the body (i.e. head, pronotum and coxal plates), but are often arranged in rows or groups. Finer, evenly dispersed punctures, in some cases only visible at more than × 50, occur in most species. This fine punctuation, or micro-punctuation, may cover almost the entire surface in some species, but in many species it is largely confined to patches on the legs. Elytra of some species are micro-punctured in the females, while males of the same species lack micro-punctuation.

The dorsal surface appears largely glabrous in the European species, the setae being extremely short.

Female genitalia (Fig. 118)

Externally comprised by the invaginated segments VIII and IX (Fig. 118). Ventral

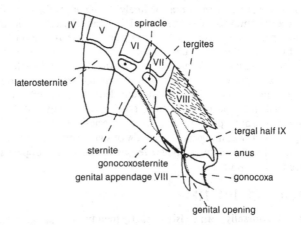

Fig. 118. Posterior part of female abdomen of *Haliplus flavicollis* Sturm; genital sclerites partly extruded, lateral view.

parts of abdominal segment VIII give rise to the gonocoxosternites (termed "S VIII" by Balfour-Browne (1950)); these are formed by a pair of triangular sclerites, each with and anteriorly directed strut. Sclerotized parts of T IX present as two anteriorly directed struts ("valvifers"), attached to the paired gonocoxae ("genital valves"); these are flattened, and in most species triangular. Gonocoxosternites, parts of T IX, gonocoxae, and the genital appendages VIII ("vulvosclerites") comprise the sclerotized parts of the ovipositor. Internal female genitalia largely membranous. Further details are given by Burmeister (1976).

Male genitalia (Fig. 119)

Externally comprised by the invaginated segment IX and perhaps non-sclerotized remnants of post-genital segments (Fig. 119). Urite IX (genital plate or gonosomite) probably consists of S IX and T IX which are fused into a sheath partly enclosing the aedeagus at rest. Aedeagus consists of two lateral parts, the parameres, and a median phallic part, the penis; parameres proximally movably attached to the penis. Aedeagus not bilaterally symmetrical. Left paramere rather short and flattened, with more sparse setation. Right paramere longer, with fringes of long hairs in most species, and in some species with a distal extension, the digitus (also seen in the left paramere of a few species). Penis asymmetrical and laterally compressed. It is long and curved with a tongue-shaped or pointed apex, and has a sperm-groove on the convex dorsal side. Distal part of penis consisting of a number of closely associated lobes (Balfour-Browne, 1915), the shape of which varies greatly from species to species.

When at rest in the abdomen, the aedeagus lies almost horizontally on its right side. During copulation the organ is turned downwards, and in this position the aedeagus

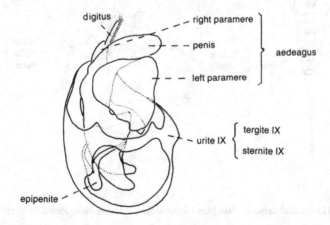

Fig. 119. Male genital sclerites of *Haliplus mucronatus* Stph., dorsal view. Redrawn from Franciscolo (1979).

has actually turned 180° compared to the primitive condition in the Coleoptera (Balfour-Browne, 1940a). Thus, what is here termed the right side of the aedeagus is truly the left side, and what is termed the dorsal side is the true ventral side of the organ.

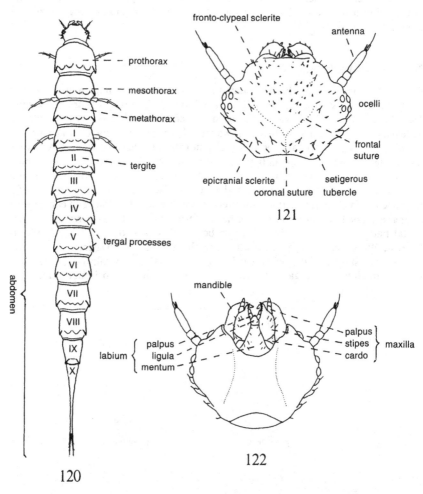

Fig. 120. Third instar larva of *Haliplus confinis* Stph., dorsal view. Redrawn from Schiødte (1864).

Figs 121, 122. Head of larva of *Haliplus* sp. – 121: dorsal view; 122: ventral view. Redrawn from Bertrand (1928).

74

Morphology of the larva

(Figs 120-123)

Body long or very long, subcylindrical, yellowish, greyish or brownish, often overgrown with diatoms.

Head (Figs 121, 122) prognathous. Head capsule dorsally with an anterior frontoclypeal sclerite which has the front margin produced into two lobes. These are partly covered by a dorsal fold. Head dorsally with two posterior epicranial sclerites. These three mentioned sclerites are separated by coronal and frontal sutures, the latter indistinct anteriorly. A neck-region is absent. Head on each side with a group of 6 ocelli. Labrum absent.

Antennae 4-segmented; the distal segment very small. Antennae often with a number of setae and short spines.

Mandibles short, wide and flattened, with a hook-shaped apex and a suction-tube formed by a partly closed groove; inner edge provided with a small tooth, in *Brychius* also with a spine-carrying lobe.

Maxillae with a small cardo and a large squarish stipes carrying the triarticulate palpus. Next to the palpus the maxilla has two lobes of which the outer one corresponds to an unsegmented galea. The maxillae are intimately connected to the labium forming a maxillolabial complex.

Labium rather small, largely composed by the mentum, distally with a small ligula and two-segmented palpi.

Buccal orifice largely open, though only the lateral parts connecting with the mandibular suction-tube are used for the uptake of nutrition.

Thorax with well sclerotized tergites covering the dorsal surface. Ventral surface largely membranous, excepting small sclerotized parts of pleurites and sternites. Mesothorax with a pair of spiracles in the third instar larva.

Legs (Fig 123) generally short, especially front legs. Each leg is composed of a coxa, trochanter, femur, tibia and a tarsus bearing one terminal claw; the trochanter sometimes appears to be two-segmented. Legs provided with a number of spines, without swimming-hairs; modified for crawling on submerged vegetation. Front femur or tibia in some species with a distal dilation ventrally which is used for holding the filamentous algae on which the larvae of these species feed.

Abdomen with 10 segments, but in *Peltodytes* yet appearing 9-segmented. Segments 1-9 (in *Peltodytes* 1-8) of subequal length, dorsally covered by well sclerotized tergites, ventrally largely membranous, the sternites being only weakly sclerotized. Segment 10 long, attenuated and bifurcate distally in most species; in *Peltodytes*, however, segments 9 and 10 are fused into one segment composed of a short conical basal part and a distal bifurcate part with a pair of very long and segmented appendages. Apical segment (10 or 9 + 10) partly membranous ventrally where the anus is situated; in *Peltodytes* with one sternal sclerite in front of anus and another behind it; in *Haliplus* with two pairs of sclerites behind anus. Most abdominal segments of the third instar larva carry a pair of spiracles in the pleural membrane.

75

The surface of head, thorax and abdomen of *Haliplus* and *Brychius* with numerous setiferous tubercles which are actually micro-tracheal gills (Seeger, 1971a); in *Peltodytes* they are scarce and mainly confined to the ventral surface.

Dorsal side of thoracic and abdominal segments, excepting the last segment, with 1-3 pairs of processes or appendages. These are present as long setiferous tubercles in first instar larvae of *Peltodytes,* and become very long, segmented and filiform (tracheal gills) in the following instars, resembling the appendages of the apical segment; in other species they may be present as more or less developed, backwardly projecting, processes of the tergites.

Thoracic and abdominal segments with sparse setation.

Larvae of the first two instars (L I and L II) with fewer setiferous tubercles than larvae of the third (and final) instar (L III). For further information on the immature stages, see Bertrand (1928, 1972) and Seeger (1971a).

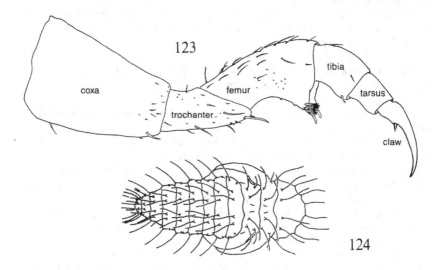

Fig. 123. Front leg of larva of *Brychius elevatus* (Panz.).
Fig. 124. Pupa of *Haliplus lineatocollis* (Marsh.), dorsal view. Redrawn from Bertrand (1928).

Zoogeography

The Haliplidae comprise about 200 described species, and the family occurs in all major faunal regions of the world. A comparatively high number of species is known from the temperate and subtropical zones, particularly in the northern hemisphere; about 50 species occurs in the Palearctic region and about 70 in the Nearctic region.

The genus *Haliplus* is almost world-wide distributed and dominating in the Palearctic region. Of the remaining Palearctic genera *Peltodytes* occurs in most parts of the world, particularly in the warmer climates, and is represented by a rather high number of species in N.America, while *Brychius* is confined to the Holarctic region. Two genera, both monotypic, do not occur in the Palearctic region (*Algophilus:* South Africa; *Apteraliplus:* western N.America).

Two species, *Haliplus apicalis* and *H. fulvus* are Holarctic, occurring in parts of N.America and with a Eurosiberian distribution in the Palearctic region; only *H. fulvus* penetrates into the Mediterranean area.

Eurosiberian species not or almost not penetrating into the Mediterranean area are *Brychius elevatus, Haliplus confinis, H. lineolatus, H. sibiricus, H. wehnckei, H. interjectus,* and *H. immaculatus.* Also Eurosiberian, but penetrating deeper into the Mediterranean area, is *Haliplus flavicollis.*

A large West Palearctic group comprises *Peltodytes caesus, Haliplus lineatocollis, H. varius, H. obliquus, H. furcatus, H. fulvicollis, H. fluviatilis, H. ruficollis, H. heydeni, H. laminatus, H. mucronatus* and *H. variegatus.* These species do not occur in the East Palearctic, although some of them may reach east of the Ural mountains. They are generally widely distributed in the Mediterranean area, and a few also occur in N.Africa. In Fennoscandia they usually have a southern or south-eastern distribution.

Bionomics

The Fennoscandian species of Haliplidae inhabit a large variety of freshwater habitats, and some species even occur in brackish water. The majority of species may be met with in clean nutrient-rich ponds and lakes, and in the backwaters of streams. Places under strong influence of current or waves or being strongly shaded, and temporary or very acid water bodies are avoided by most species. Seeger (1971a, b, c) provides many ecological facts about species of *Haliplus.*

Adult haliplids are not particularly good swimmers, partly because they move their hind legs alternately while swimming, like the hygrobiids. They spend most of the time crawling on submerged vegetation. The species vary from being almost exclusively carnivorous to being largely herbivorous in the adult. The diet may be highly specific as in *Haliplus lineolatus* which prefers freshwater hydrozoans (Seeger, 1971b). The herbivorous diet is largely composed of green algae (filamentous algae or characeans). Some haliplids probably often prefer a particular species of algae, but not much is known on this subject.

Below the surface an air supply is carried along under the elytra and under the hind coxal plates; this supply must be renewed from the surface from time to time.

Most Fennoscandian species of Haliplidae are capable of flight, but with the exception of *Haliplus lineatocollis* they only rarely fly. Jackson (1952a, b, 1956) found that the development of certain muscles and sclerites necessary for flight varies intraspecifically in many species, and that the development of flight musculature probably varies

77

through the life-span of some specimens. Species of Haliplidae are only rarely attracted to light (Jackson, 1973).

The eggs of our Haliplidae are laid in the spring or summer. Species of *Haliplus* deposit their eggs inside or between cells of aquatic plants; at least in some cases the adult has first consumed the content of the plant cell (Beier, 1929). The egg is elongate or oval, of whitish colour. In species of *Haliplus* it is rather small, about 0.35-0.45 mm. long, and lacks surface sculpture (Seeger, 1971a). The number of eggs produced by one female varies between 25 and 60 (Seeger, l.c.). The incubation period is dependent on temperature and lasts from 8 to 16 days (Hickman, 1931; Seeger, l.c.).

The larvae usually inhabit the same habitat as the adults. They crawl about on the vegetation or the bottom substratum, to which they cling by help of the legs. When disturbed they often coil around the substratum. Larvae of *Peltodytes* are rather agile, while larvae of *Brychius* and *Haliplus* are more sluggish.

The haliplid larvae feed on green algae. Larvae of *Haliplus* subgenera *Haliplidius* and *Liaphlus* (excepting *H. laminatus*) chiefly feed on characeans on which they crawl, crushing the cell walls with the mandibles, and sucking out the content. Larvae of the remaining species feed on filamentous algae; the algal filament is held by the front legs and sucked out through the mandibles.

The larvae do not need to come to the surface to respire, as gas-exchange takes place through the cuticle, probably mainly by tracheal or micro-tracheal gills.

The time needed for larval development is dependent on temperature, and may vary considerably due to climatic conditions, even within the same species (Seeger, 1971a). In nature larval development probably never lasts more than a year. Larvae of many species are capable of hibernating, either in the water or ashore. In particular, third instar larvae often hibernate ashore, pupating the following spring without re-entering the water. Larvae of other species pupate in the summer, without hibernating (Holmen, 1981).

Before pupating the larva leaves the water, buries itself in the soil ashore, and constructs a pupal cavity. Here it may rest for more than a month before pupation takes place. During this period respiration through the spiracles begins.

The pupa (Fig. 124) is yellowish white, slightly shorter than the adult. It is distinguished from pupae of other aquatic Coleoptera by the presence of dorsal setiferous tubercles, and by the absence of urogomphi (urogomphi are also absent in the gyrinid pupa). A key to some of the Fennoscandian genera and subgenera is provided by Bertrand (1972), and Seeger (1971a) gives information on the pupae of some species of *Haliplus*.

The pupa rests on its dorsal side, separated from the substratum by dorsal setae. The time spent in the pupal stage is longer for large species than for small species, and varies between 9 and 13 days at 20°C in species examined by Seeger (l.c.).

The newly hatched adult rests for about 5 to 9 days (20°C) in the pupal cavity before re-entering the water (Seeger, l.c.).

The young adults of many species do not reproduce until the following year (Seeger, l.c.). The adult may live for at least 2 years; in most species it prefers to hibernate out of the water, buried in soil or litter ashore.

78

The ectoparasitic fungus *Hydraeomyces halipli* (Thaxter) has been found in Finnish specimens of *Haliplus fulvus, H. fulvicollis* and *H. lineolatus* (Huldén, 1983b).

Collecting

Adults and larvae are best collected in suitable habitats using a strong entomological net with a mesh size of 1-2 mm. The size of the net mainly depends on the depth of the habitat being investigated, as the insects move fairly slowly; even a kitchen sieve is often very useful. Hauling through filamentous algae or characeans usually provides specimens of Haliplidae.

Adults can be searched for in the field by examining the content of the net in a light coloured tray with water-covered bottom. Larvae are very easily overlooked in the field, but may be obtained in the following way: the content of the net is placed in a sieve (mesh size about 2 mm.) on a light coloured tray with water covered bottom. When the material in the sieve successively dries out and decays, the larvae will try to escape and are collected in the tray. The tray should be examined for larvae regularly over a couple of days, and predaceous insects should be removed.

Larvae and adults often hibernate ashore, near the edge of the water. Here they can be collected by sifting the soil, leaf litter or moss in which they bury, or by turning stones and pieces of wood. During summer adults, larvae and pupae can be found in similar places, resting in the pupal cavities.

Flying adults are very rarely met with in light traps.

Rearing

Several authors have studied species of Haliplidae in captivity, and much information on the biology and morphology of various developmental stages has been provided in this way. However, the immature stages of many species are still unknown, but could be obtained by rearing *ex ovo*. Seeger (1971a, b) used the following method to rear a number of species:

Adults, eggs and larvae are kept in plastic bowls of 20×20×6 cm. Each bowl should be two thirds filled with water, and should contain a bottom layer of clean sand and a few stones. Green algae (filamentous or characeans) are provided as nutrition for larvae and adults, and for the deposition of eggs. Pieces of *Asellus* or chironomid larvae, as well as living chironomid larvae, may be supplied to adults of carnivorous species. Some species have a rather specialized diet, both as larvae and adults, and it may be necessary to try a number of different food items.

Full grown third instar larvae are transferred to another bowl of similar size. This bowl only contains a small amount of water, and it is placed in an oblique fashion, so that the water only covers the bottom of the lower quarter of the bowl. The water should contain green algae as nutrition for the larvae. A thick layer of moist sand covers the upper (dry) half of the bottom, and a thin layer of leaf litter is placed on the

79

bottom of the bowl between the sand and the water. The full grown larvae eventually leave the water and pupate in the sand.

It should be noted, that the life-cycle of haliplids is generally much faster in the laboratory, than under natural, colder, conditions.

Killing and mounting

The methods are the same as described previously for the Gyrinidae. However, extrusion and preparation of the genitalia are more difficult. This is due to the small size, irregular shape, and often weaker sclerotization of the haliplid genital sclerites. The examination of male genital sclerites is facilitated if the penis and the parameres are separated. Female genitalia are best examined on permanent slides.

The weakly sclerotized or membranous parts of the female genitalia, such as the spermatheca and the ductus receptaculi, are liable to provide good diagnostic characters. However, these structures have only been studied in very few species.

Nomenclature

The generic names used here are those which have been generally used since the catalogue on the family by Zimmermann (1920). The subgeneric names follow Guignot (1928) for *Haliplus* and Satô (1963) for *Peltodytes*. Other recent European authors have usually followed Guignot (1955) for *Haliplus,* but this work is based on an incorrectly fixed type species of the genus (see note under *Haliplus*). The reason for the inclusion of subgeneric names in this work, is that future workers on the family well raise some of the subgenera to genera, the present generic subdivision of the family appearing rather unsatisfactory from a phylogenetic point of view (genus *Haliplus* s.lat. is probably not a monophyletic unit). Furthermore, the subgeneric names are used in many important works on the family. However, revisional work on this no doubt monophyletic family is strongly needed.

The names used by Silfverberg (1979) are largely followed here at the species level, but the separation into subspecies is avoided; in my opinion, the cases previously treated in the Nordic countries as subspecies represent clinal variation. The changes in synonymy are explained under the respective taxa. Some synonyms not important to our fauna have been omitted, as have some uncertain synonyms.

Lectotype and paralectotypes have been designated for a number of taxa belonging to a group of closely related species within subgenus *Haliplus* s.str. A revision of this group is in preparation (Holmen, in prep.).

Key to genera of Haliplidae

Adults

1   Pronotum almost square, with the widest point near the
middle (Figs 125a-127a). Elytra with longitudinal ridges (Figs

80

125b-127b). Length: 3.5-4.3 mm................ *Brychius* Thomson (p.84)
- Pronotum widest at the base, with sides converging anteriorly
  (Fig. 107). Elytra without longitudinal ridges......................... 2
2(1) Elytra with a fine presutural groove in apical half (Fig. 128).
  Hind coxal plates bordered laterally (Fig. 129). Length: 3.5-
  4.0 mm .... *Peltodytes* Régimbart (p.86)
- Elytra without a presutural groove. Hind coxal plates not bor-
  dered. Length: 2.0-4.6 mm .... *Haliplus* Latreille (p.89)

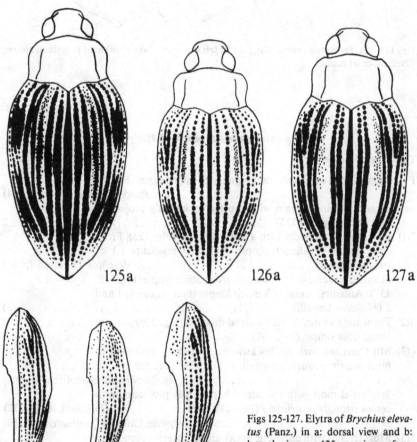

Figs 125-127. Elytra of *Brychius eleva-tus* (Panz.) in a: dorsal view and b: lateral view. – 125: specimen from southwestern Denmark; 126: specimen from southeastern Norway; 127: specimen from northeastern Finland.

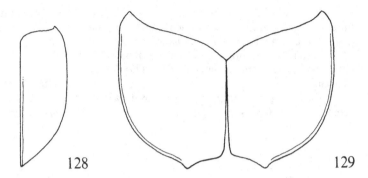

Figs 128, 129. *Peltodytes caesus* (Dft.). – 128: left elytron to show position of presutural groove; 129: hind coxal plates.

Key to genera and subgenera of Haliplidae

Third instar larvae

1   Tergites with extremely long segmented appendages (Fig. 130) . . . .
    . . . . *Peltodytes* Régimbart (p.86)
–   Tergites at hind margin with shorter backwardly projecting
    processes (Figs 131-135) 2

2(1) Front femur ventrally with a distal dilation (Fig. 123). Front
    tibia simple or almost simple. Antennal segments 1-3 of
    subequal length . . . . *Brychius* Thomson (p.84)
–   Front femur, and sometimes also front tibia, simple (Figs 136,
    137). Antennal segment 3 much longer than segments 1 and
    2 (*Haliplus* Latreille) 3

3(2) Front tibia ventrally with a distal dilation (Fig. 136) 4
–   Front tibia simple (Fig. 137) 5

4(3) Most tergites with all backwardly projecting processes at
    hind margin strongly protruding (Fig. 131) . . . .
    . . . . sg. *Neohaliplus* Netolitzky (p.108)
–   Tergites at most with the lateral backwardly projecting pro-
    cesses strongly protruding (Figs 132, 133) . . . . sg. *Haliplus* s.str. (p.122)
                  or sg. *Liaphlus* Guignot (*laminatus*) (p.110)

5(3) Most tergites with only four backwardly projecting processes
    distinct (Fig. 134) . . . . sg. *Liaphlus* Guignot (p.110)
–   Most tergites with six backwardly projecting processes dis-
    tinct (Fig. 135) . . . . sg. *Haplidius* Guignot (p.117)

82

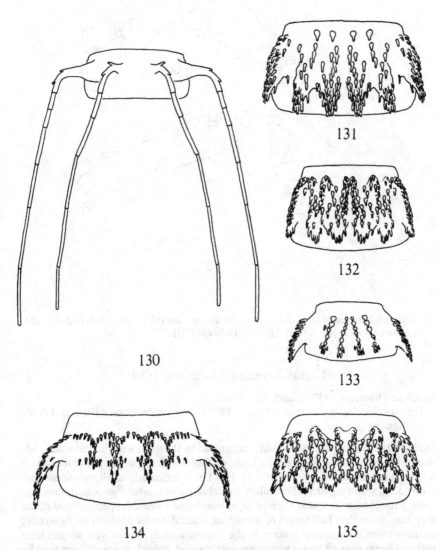

130

131

132

133

134

135

Figs 130-135. Abdominal segment I of third instar larvae of Haliplidae, dorsal view. – 130: *Pelto-dytes caesus* (Dft.); 131: *Haliplus lineatocollis* (Marsh.); 132: *H. immaculatus* Gerh.; 133: *H. ful-vicollis* Er.; 134: *H. variegatus* Sturm; 135: *H. confinis* Stph. – 131, 132, and 135 redrawn from Seeger (1971a).

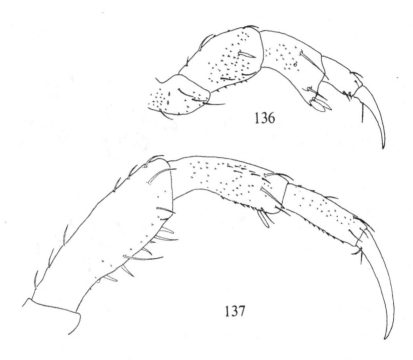

Figs 136, 137. Front leg of larva of 136: *Haliplus lineatocollis* (Marsh.) and 137: *H. fulvus* (F.). – 136 redrawn from Seeger (1971a), 137 from Holmen (1981).

Genus *Brychius* Thomson, 1859

Brychius Thomson, 1859, Skand. Col. 1: 11.
   Type species: *Dytiscus elevatus* Panzer, 1794 [= *Brychius elevatus* (Panzer, 1794)], by monotypy.

Medium-sized or large species, easily recognized by the elongate, discontinuous outline of the body. Most parts of the body surface with both larger punctures and a micro-punctuation visible at about ×50. Head not particularly small, eyes somewhat protruding; apical segment of maxillary and labial palps shorter than penultimate segment. Pronotum widest near the middle, at least in the Palearctic species; basal plicae very long, more than half length of pronotum. Lateral border of elytra finely serrate; elytra without a presutural groove; in the Fennoscandian species with longitudinal ridges. Elytral rows of larger punctures well marked, though some of them may be strongly abridged or confluent. Elytral epipleura very wide apically. Hind coxal plates not reaching beyond sternite V, not bordered laterally. Hind tibia with a setiferous striole of somewhat scattered punctures. Sternite VII with a longitudinal groove in the middle. Male generally slightly smaller than female, with front and mid tarsal segments 1-2 or 1-3 dilated and provided with tufts of sucker-hairs on the ventral side.

Four Nearctic and two Palearctic species are generally recognized. The only Fennoscandian species forms a number of local races, some of which have been treated as distinct species by various authors.

Members of this genus clearly prefer running water, although they are occasionally met with in lakes.

14. *Brychius elevatus* (Panzer, 1794)
    Figs 123, 125-127, 138-142; pl. 1: 2.

Dytiscus elevatus Panzer, 1794, Faun. Germ. 1 (14): t. 9.
Brychius cristatus J. Sahlberg, 1875, Notis. Sällsk. Faun. Fl. fenn. Förh. 14: 137.
Brychius rossicus Semenov, 1898, Horae Soc. Ent. Ross. 31: 542.
Brychius elevatus ssp. *norvegicus* Munster, 1922, Videnskapselsk. Skr. 1 (9): 6.

3.5-4.3 mm. Dorsal side yellowish, often infuscated along the posterior margin of the head, and along the anterior and posterior margins of pronotum; elytra with black longitudinal lines (Figs 125a-127a). Ventral side, including legs, yellowish or reddish. Pronotum almost square, with basal plicae almost reaching the anterior margin. Elytral rows of large punctures well marked, rows 5 and 6 (counted from the suture, presutural fine row not included) confluent and only present anteriorly. The anterior two thirds of each elytron with a longitudinal ridge situated between rows of punctures

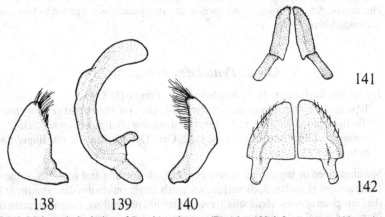

141

142

138   139   140

Figs 138-140. Male genital sclerites of *Brychius elevatus* (Panz.). – 138: left paramere; 139: penis; 140: right paramere.
Figs 141, 142. Female genital sclerites of *Brychius elevatus* (Panz.). – 141: gonocoxae and tergal halves IX; 142: gonocoxosternites.

85

2 and 3, and sometimes with a small basal elevation between rows 4 and 5 (Figs 125b-127b). Prosternal apophysis with bordered sides. Metasternal apophysis with a large median impression. Male: front claws simple, of subequal length; penis with a wide, rounded apex; both parameres with an apical fringe of hairs along the ventral edge; genital sclerites, vide Fig. 138-140. Female: gonocoxosternites large, triangular, with short struts; tergal halves IX short and narrow; gonocoxae oblong, parallel-sided, rounded apically; genital sclerites, vide Figs 141, 142.

Distribution. Denmark: rather common in Jutland; not east of F. – Sweden: scattered records throughout the country; often common locally. – Norway: only in AK and HEs in the south and TRi, Fi and Fø in the north. – Finland: recorded from almost all districts; also adjacent parts of the USSR. – Most of Europe except southern and western parts, but present in Great Britain; northern and central parts of the USSR, east to central Siberia.

Biology. In streams and rivers with sandy or gravelly bottom; occasionally in lakes. The larva, and perhaps also the adult, feeds on filamentous algae. Not much is known about its life-cycle. Larvae have been collected in the summer and autumn, and teneral adults in the autumn, so the larvae probably do not hibernate. Adults hibernate in the water. The larva was first described by Rousseau (1920).

Note. The shape of the elytral ridges varies somewhat. Specimens from southern populations generally have the ridges low, particularly in the anterior part (*"elevatus*-form"; Fig. 125b), while northern specimens have the ridges more strongly developed ("*cristatus*-form"; Fig. 127b). Transitional forms (Fig. 126b) occur in northern Denmark and in some southern populations in Sweden, Norway and Finland. Local populations often differ slightly from each other with respect to a number of morphological characters. A subdivision of this species into subspecies, as proposed by some authors, is avoided here.

Genus *Peltodytes* Régimbart, 1878

Peltodytes Régimbart, 1878, Annls Soc. ent. France (5) 8: 450.
Type species: *Dytiscus caesus* Duftschmid, 1805 [= *Peltodytes (Peltodytes) caesus* (Duftschmid, 1805)], by subsequent designation (Balfour-Browne, 1936).
Cnemidotus Illiger sensu Erichson, 1832, Gen. Dyticeorum: 19; *nec* Illiger, 1802; see note under *Haliplus*.

Medium-sized or large species, very convex both dorsally and ventrally, some species being almost globular. Body surface with only larger punctures discernable, at least in the European species. Head small, eyes somewhat protruding. Apical segment of maxillary and labial palps longer than the penultimate segment. Pronotum narrowed anteriorly from the base, without basal plicae. Lateral border of elytra not serrate; elytra with a fine groove along the apical half of the suture, in our species without dorsal ridges or elevations; punctures very large, arranged in rows which are often irregular,

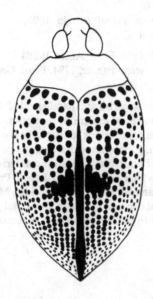

Fig. 143. *Peltodytes caesus* (Dft.), dorsal view.

abridged or confluent. Elytral epipleura rather narrow apically. Hind coxal plates large, almost reaching sternite VII, bordered laterally. Hind tibia with a distinct setiferous striole. Sternite VII without a longitudinal groove in the middle. Male with segments 1-2 of front and mid tarsi dilated and provided with tufts of sucker-hairs on the ventral side.

Most of the approximately 40 known species of this genus are found in North America; a few species live in the northern parts of the Neotropical region, and a few are known from the Palearctic, Afrotropical and Oriental regions. Several more species probably occur in the tropics. Two species occur in Europe, only one in the north. Two subgenera have been recognized: *Peltodytes* s. str. and *Neopeltodytes* Satô, 1963.

Members of this genus prefer stagnant bodies of water with rich vegetation.

Subgenus *Peltodytes* s. str.

Peltodytes sg. *Peltodytes* s. str.

This subgenus is recognized on a single character: hind coxal plate with a posterior denticle (Fig. 129).

The Palearctic, Oriental and Afrotropical members of *Peltodytes* all belong to this subgenus.

15. *Peltodytes (Peltodytes) caesus* (Duftschmid, 1805)
Figs 128-130, 143-148; pl. 1: 1.

Dytiscus caesus Duftschmid, 1805, Faun. Austr. 1: 284.
Dytiscus impressus Fabricius sensu Panzer, 1794, Faun. Germ. 1 (14): t. 7; misident., *nec* Fabricius, 1787.

3.5-4.0 mm. Body wide, rather squarish and parallel-sided. Dorsal side yellowish with black punctures along the posterior margin of pronotum and on elytra, and with a dark common spot and often some minor spots on the elytra (Fig. 143). Ventral side and legs yellowish or reddish. Pronotum with a coarsely punctured impression along the posterior margin. Elytral punctures very strong anteriorly, particularly the first punctures of the inner rows which are almost pit-shaped. Prosternal apophysis with a median depression apically, and with slightly raised sides. Metasternal apophysis with a median pit and some lateral depressions. Each hind coxal plate with a blunt denticle posteriorly (Fig. 129). Male: front claws simple, of subequal length; penis stout with

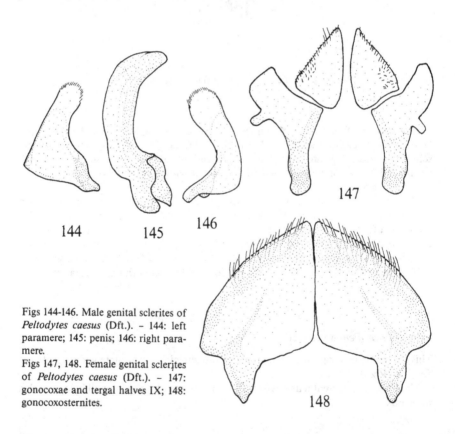

144   145   146   147

Figs 144-146. Male genital sclerites of *Peltodytes caesus* (Dft.). – 144: left paramere; 145: penis; 146: right paramere.
Figs 147, 148. Female genital sclerites of *Peltodytes caesus* (Dft.). – 147: gonocoxae and tergal halves IX; 148: gonocoxosternites.

148

a blunt apex; parameres with an apical membraneous part, bordered with short hairs; genital sclerites, vide Figs 144-146. Female: gonocoxosternites large, triangular, with short struts; gonocoxae small, triangular; tergal halves IX wide, weakly sclerotized; genital sclerites, vide Figs 147, 148.

Distribution. In the southeast of Denmark, not in WJ, NWJ and NEJ. – In the south of Sweden: Sk., Sm. and Öl. – Not in Norway or East Fennoscandia. – Most of Europe, north to Britain and Fennoscandia; central and southern parts of European USSR, northern Africa, Israel, Syria, Asia Minor, Iran, Transcaucasia, Turkmenia and Afghanistan.

Biology. Mainly in stagnant, eutrophic bodies of water with rich vegetation. The larva, and probably also the adult, feeds on filamentous algae. Eggs are laid in the spring on the surface of submerged vegetation. Full-grown larvae occur later in the summer, as well as teneral adults, so the larva probably does not hibernate. The larva was first described by Schiødte (1872), and the pupa by Bertrand (1928). In Fennoscandia the adult probably hibernates out of the water.

Genus *Haliplus* Latreille, 1802

Haliplus Latreille, 1802, Hist. Nat. Crust. Ins. 3: 77.
Type species: *Dytiscus impressus* Fabricius, 1787 [= *Haliplus (Haliplus) ruficollis* (De Geer, 1774)], by subsequent designation (Latreille, 1810); see note.
Cnemidotus Illiger, 1802, Mag. f. Insektenkunde 1: 297.
Type species: not designated; see note.
Hoplitus Clairville, 1806, Ent. Helv. 2: 218.
Type species: not designated; see note.
Halipus, sensu auctt.; incorrect spelling.

Small, medium-sized or large species. Body usually very convex both dorsally and ventrally, and with a rather continuous outline. Most of body surface with large punctures of varying sizes. Some species have micro-punctuation on the dorsal surface; it is discernable at about ×20-50 and may cover almost the entire dorsal surface (sometimes also the ventral surface), or may be confined to limited parts (e.g. female elytra). Head not particularly small in our species, eyes somewhat protruding; apical segment of maxillary and labial palps shorter than penultimate segment. Pronotum narrowed from or almost from base; basal plicae often present. Lateral border of elytra not serrate in the Fennoscandian species; elytra neither with a presutural groove nor with dorsal ridges or elevations. Each elytron with large punctures forming about 10 distinct rows; often also with scattered punctures and more or less irregular, often sparse rows of finer punctures between the main rows and presuturally. Elytral epipleura rather narrow distally. Hind coxal plates not reaching beyond sternite V, not bordered laterally. Hind tibia with or without a setiferous striole. Sternite VII without a longitudinal groove in the middle. In males segments 1-3 of front tarsi and 1 or 1-3 of mid tarsi are dilated, with tufts of sucker-hairs on the ventral side. Males in some species slightly longer and narrower than females. Elytra in some species micro-punctured in females and smooth in males.

About 130 species are known. The genus is distributed throughout most of the world, with a large number of species in the Nearctic region, and 20 species in northern Europe. At present the genus is divided into 6 subgenera of which four occur in Europe: *Haliplus* s. str., *Neohaliplus* Netolitzky, 1911, *Haliplidius* Guignot, 1928 and *Liaphlus* Guignot, 1928.

Members of the genus occur in a large variety of aquatic habitats, but the majority are found in clear, base-rich stagnant bodies of water.

Notes on synonymy. Latreille (1802) created genus *Haliplus* for two species, viz., *Dytiscus impressus* Fabricius, 1787 sensu Latreille [misident., = *Dytiscus fulvus* Fabricius, 1801] and *Dytiscus obliquus* Fabricius, 1787; subsequently Latreille (1810) designated *Dytiscus impressus* Fabricius, 1787, as type species of his genus.

Latreille (1806) figured his *impressus,* and it is evident that his descriptions (Latreille, 1802, 1806) were based on specimens of the species now known as *Haliplus fulvus* (Fabricius, 1801). A lectotype of *Dytiscus impressus* Fabricius, 1787 [= *Dytiscus ruficollis* De Geer, 1774] is designated below under *Haliplus ruficollis.* This designation is in accordance with the synonymy presented by most authors, which has furthermore led some authors to consider *Dytiscus impressus* Fabricius, 1787 to be the type species of *Haliplus* Latreille, 1802.

Whether *Dytiscus impressus* Fabricius, 1787 or *Dytiscus impressus* Fabricius, 1787 sensu Latreille is to be considered the type species of *Haliplus* Latreille, 1802, should be decided by the International Commission on Zoological Nomenclature. For the time being, I prefer to follow the former view, as it best supports the stability of the nomenclature presently used. This view is also followed by most recent N. American authors.

European authors largely follow Balfour-Browne (1938), who proposed *Dytiscus obliquus* Fabricius, 1787, as the type species. His designation is, however, clearly not valid according to the ruling by the ICZN (art. 69, a, vi). The subgeneric names used below are therefore not in accordance with most recent European works on the genus (*ruficollis, fulvus* and *obliquus* all belong to different subgenera).

Illiger (1802) created the genus *Cnemidotus* for the following species: *Dytiscus impressus* Fabricius, 1787, *D. obliquus* Fabricius, 1787, and *D. elevatus* Panzer, 1794. Hope (1838) proposed *Dytiscus caesus* Duftschmid, 1805 as type species of the genus, but his fixation is not valid according to the ruling by the ICZN (art. 69, a, iv) as this name was not synonymised with any of Illiger's species. At present no type species seems to have been validly fixed for *Cnemidotus* Illiger, 1802. Illiger (1798, 1802) no doubt identified his *obliquus* and *elevatus* correctly, but apparently included all other species of Haliplidae in his *impressus.*

Clairville (1806) named the genus *Hoplitus* for four species, viz., *Dytiscus fulvus* Fabricius, 1801, *Dytiscus impressus* Fabricius, 1787, *Dytiscus obliquus* Fabricius, 1787 and *Dytiscus marginepunctatus* Panzer, 1794. No type species has been fixed for the

90

genus. The identity of Clairville's species is too uncertain at present to state which species would best qualify as the type species of *Hoplitus* Clairville, 1806.

Key to species of *Haliplus*

1   Body surface almost completely covered by a coarse micro-punctuation, discernable at about ×20 (Fig. 112). Larger punctures rather weakly impressed. (Sg. *Haliplidius* Guignot) 2

–   At least pronotum largely without micro-punctuation, when present, discernable at about ×50 (Figs 113, 115). Larger punctures more deeply impressed 4

2(1)   Prosternal apophysis distinctly bordered along the sides (Fig. 149). Pronotal basal plicae distinct (Fig. 178). Punctures on head of subequal size. Length: 3.0-3.8 mm.. 21. *confinis* Stephens

–   Prosternal apophysis not bordered along the sides. Pronotal basal plicae weak or absent. Punctures on head of two sizes 3

3(2)   Anterior and posterior margins of pronotum black (Fig. 165). Femora with a very narrow black ring distally. The four dark lines closest to the suture of each elytron hardly interrupted (Fig. 165). Pronotal basal plicae very weak (Fig. 179). Length: 2.5-3.0 mm .... 22. *varius* Nicolai

–   Anterior and posterior margins of pronotum infuscated, but not black. Femora without a black ring distally. The four dark lines closest to the suture of each elytron usually widely interrupted (Fig. 166). Pronotal basal plicae absent. Length: 3.1-3.8 mm .... 23. *obliquus* (Fabricius)

4(1)   Pronotum without basal plicae. Upper face of hind tibia with a setiferous striole (Figs 186-188). (Sg. *Liaphlus* Guignot) 5

–   Pronotum with basal plicae (Figs 180-185). Upper face of hind tibia without a setiferous striole (Figs 189-191) 9

5(4)   Sides of prosternal apophysis extending forwards as plicae to the anterior edge of prosternum (Fig. 150) 6

–   Sides of prosternal apophysis not extending forwards as plicae to the anterior edge of prosternum (Fig. 151). Length: 3.4-4.1 mm .... 18. *flavicollis* Sturm

6(5)   Head very wide, the distance between the eyes at least twice the maximum width of one eye (Fig. 160). Elytra without dark blotches in addition to dark punctures. Length: 3.7-4.3 mm .... *mucronatus* Stephens

–   Head narrower, the distance between the eyes less than

91

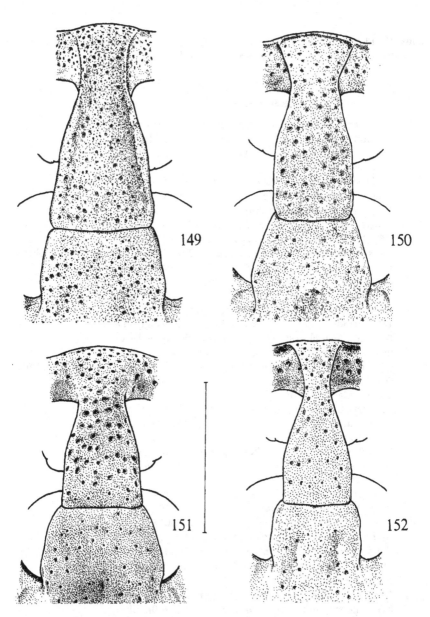

Figs 149-152. Pro- and metasternal apophyses of *Haliplus* spp. – 149: *H. confinis* Stph.; 150: *H. fulvus* (F.); 151: *H. flavicollis* Sturm; 152: *H. fulvicollis* Er. – Michael Hansen del., drawings based on electroscan micrographs. Scale: 0.5 mm for 149 and 152, 0.7 mm for 150 and 151.

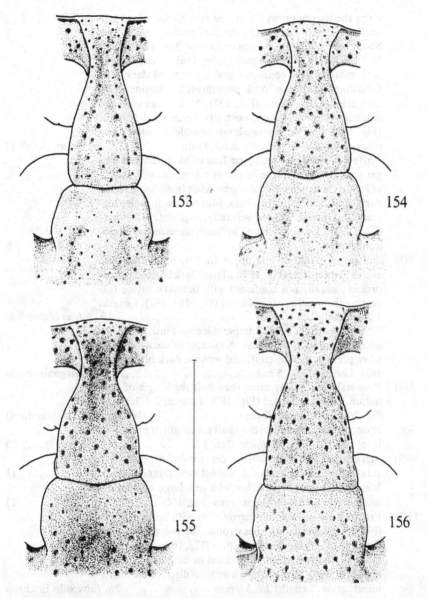

Figs 153-156. Pro- and metasternal apophyses of *Haliplus* spp. – 153: *H. furcatus* Seidl.; 154: *H. apicalis* Thoms.; 155: *H. ruficollis* (De Geer); 156: *H. fluviatilis* Aubé. – Michael Hansen del., drawings based on electroscan micrographs. Scale: 0.5 mm.

twice the maximum width of one eye. Elytra often with
dark blotches in addition to the dark punctures 7

7(6) Setiferous striole on the upper face of hind tibia about
half the length of the segment (Fig. 186). Elytra usually
with more or less confluent and interrupted dark lines
following the rows of dark punctures, but distinct blotches are usually absent (Fig. 158). Mid tarsus of male:
segment 1 strongly dilated ventrally, segments 2-5 simple
(Fig. 192). Female: elytra almost completely covered by
micro-punctuation. Length: 2.5-3.3 mm. . . . . 17. *laminatus* (Schaller)

– Setiferous striole on the upper face of hind tibia not longer than one third of the length of the segment (Figs 187,
188). Elytra usually with dark blotches in addition to the
dark punctures (Figs 161-163). Mid tarsus of male: segments 1-3 weakly dilated ventrally, segments 4-5 simple
(Fig. 193). Female: elytra without discernable micro-
punctuation 8

8(7) Setiferous striole on the upper face of hind tibia with
about 10 punctures (Fig. 187). Elytral dark blotches, when
present, usually not confluent with the dark suture (excepting the small preapical spot) (Figs 161, 162). Length:
3.5-4.6 mm . . . . 19. *fulvus* (Fabricius)

– Setiferous striole on the upper face of hind tibia with
about 5 punctures (Fig. 188). A number of dark blotches,
when present, usually confluent with the dark suture (Fig.
163). Length: 2.5-3.5 mm . . . . 20. *variegatus* Sturm

9(4) Pronotal basal plicae more than half the length of pronotum, strongly curved (Fig. 180). Length: 2.3-3.3 mm.
(Sg. *Neohaliplus* Netolitzky) . . . . 16. *lineatocollis* (Marsham)

– Pronotal basal plicae less than half the length of pronotum
(Figs 181-185). (Sg. *Haliplus* s. str.) 10

10(9) Metasternal apophysis with a depression or pit to each
side of the middle, otherwise almost even (Figs 152-154) 11

– Metasternal apophysis either with one large median de-
pression (Fig. 155) or almost even (Fig. 156) 13

11(10) Prosternal apophysis not furrowed or with bordered sides
(Fig. 152). Elytral dark lines strongly interrupted and confluent, forming distinct dark spots (Fig. 169). Male: penis
gradually narrowed into a pointed or narrowly rounded
apex (Fig. 196). Female: elytra without discernable micro-
punctuation. Length: 2.6-3.0 mm . . . . 26. *fulvicollis* Erichson

– Prosternal apophysis with a shallow median furrow basally; this furrow may branch into two, bordering the sides of the apophysis (Figs 153, 154). Elytral dark lines less

strongly interrupted and confluent (Figs 167, 168). Male: penis with a rather wide, truncate apex (Figs 194, 195). Female: elytra sometimes micro-punctured 12

12(11) Sides of prosternal apophysis distinctly bordered, also apically (Fig. 153). Elytral dark lines interrupted and often confluent behind the middle, giving the elytra a more spotted appearance (Fig. 168). Male: dorsal side of penis with a long, shallow excision (Fig. 195). Female: elytra sometimes micro-punctured, especially towards apex. Length: 2.2-2.8 mm .... 25. *furcatus* Seidlitz

  – Sides of prosternal apophysis weakly bordered basally, not bordered apically (Fig. 154). Elytral dark lines sometimes abridged at base and apex, but rarely interrupted or confluent along their course, giving the elytra a striped appearance (Fig. 167). Male: dorsal side of penis rather evenly curved (Fig. 194). Female: elytra completely covered by micro-punctuation. Length: 2.5-2.8 mm ... 24. *apicalis* Thomson

13(10) Metasternal apophysis with a large median depression (Fig. 155) 16

  – Metasternal apophysis almost even, at most with a very small median depression (Fig. 156) 14

14(13) Body rather short, widest just behind pronotum, sides converging rather strongly behind (Fig. 176). Male: penis short, not evenly curved (Fig. 203). Female: elytral micropunctuation at most confined to very narrow zones along the lateral border and the suture in the apical portion of the elytra. Length: 2.2-2.8 mm .... 33. *heydeni* Wehncke

  – Body longer; widest point not just behind pronotum (Figs 167, 170). Male: penis long, rather evenly curved (Figs 194, 197). Female: elytra micro-punctured in at least the apical two thirds 15

15(14) Elytral dark lines sometimes abridged at base and apex, but rarely interrupted or confluent along their course, giving the elytra a striped appearance (Fig. 167). Pronotal basal plicae fairly long (Fig. 181). Male: ventral side of right paramere not dilated subapically (Fig. 194). Length: 2.5-2.8 mm .... 24. *apicalis* Thomson

  – Elytral dark lines usually strongly interrupted, giving the elytra a spotted appearance (Fig. 170). Pronotal basal plicae short (Fig. 182). Male: ventral side of right paramere dilated subapically (Fig. 197). Length: 2.5-3.2 mm .... 27. *fluviatilis* Aubé

16(13) Males. Segments 1-3 of front and mid tarsi dilated ventrally, provided with tufts of sucker-hairs (Figs 205-212) 17

- Females. Segments 1-3 of front and mid tarsi simple (Figs 213-214) 23
17(16) Segment 1 of front tarsi very wide basally, with a longitudinal keel on the ventral side (Fig. 212). Elytral dark lines only interrupted and confluent to a small extent,

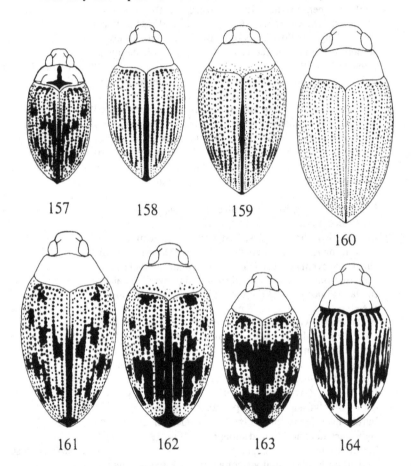

Figs 157-177. *Haliplus* spp. in dorsal view. – 157: *H. lineatocollis* (Marsh.); 158: *H. laminatus* (Schall.); 159: *H. flavicollis* Sturm; 160: *H. mucronatus* Stph.; 161: *H. fulvus* (F.), specimen from Denmark; 162: same species, specimen from northern Sweden; 163: *H. variegatus* Sturm; 164: *H. confinis* Stph.; 165: *H. varius* Nicol.; 166: *H. obliquus* (F.); 167: *H. apicalis* Thoms.; 168: *H. furcatus* Seidl.; 169: *H. fulvicollis* Er.; 170: *H. fluviatilis* Aubé; 171: *H. lineolatus* Mann.; 172: *H. interjectus* Lindbg.; 173: *H. sibiricus* Motsch.; 174: *H. wehnckei* Gerh.; 175: *H. ruficollis* (De Geer); 176: *H. heydeni* Wehncke; 177: *H. immaculatus* Gerh.

96

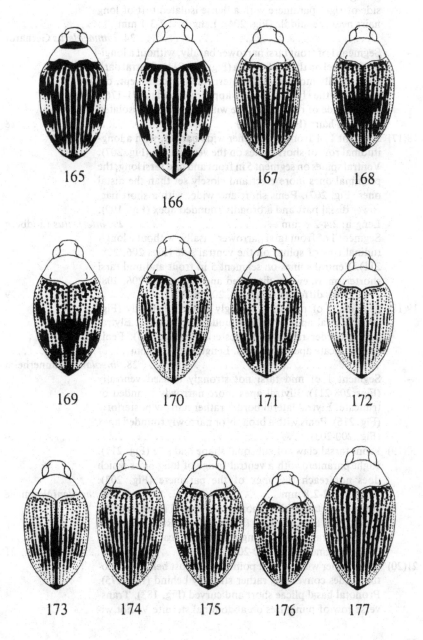

165

166

167

168

169

170

171

172

173

174

175

176

177

giving the elytra a striped appearance (Fig. 177). Ventral
side of right paramere with a dense isolated tuft of long
hairs near the middle (Fig. 204). Length: 2.7-3.1 mm . . . .
. . . . 34. *immaculatus* Gerhardt
- Segment 1 of front tarsi narrower basally, without a longi-
tudinal keel on the ventral side (Figs 205-211). Elytral dark
lines usually more strongly interrupted or confluent, often giving the elytra a spotted appearance (Figs 171-176).
Ventral side of right paramere with a longer, less isolated
fringe of hairs (Figs 198-203) 18
18(17) Segment 1 of front tarsi rather wide basally, with a long-
itudinal row of short spines on the ventral side (Fig. 207).
Ventral spines on segment 5 in front and mid tarsi long, the
proximal ones more erect and closely set than the distal
ones (Fig. 207). Penis short and wide, with a short narrower distal part and a broadly rounded apex (Fig. 199).
Length: 2.4-2.8 mm . . . . 29. *interjectus* Lindberg
- Segment 1 of front tarsi narrower basally, without a longi-
tudinal row of spines on the ventral side (Figs 206, 208-
211). Ventral spines on segment 5 in front and mid tarsi
shorter, more evenly dispersed and erect (Figs 206, 208-
211). Penis different (Figs 198, 200-203) 19
19(18) Segment 1 of mid tarsi strongly excised ventrally (Fig.
206). Elytral apex broadly rounded or truncate. Elytral
lateral border rather wide posteriorly (Fig. 215). Penis
with truncate apex (Fig. 198). Length: 2.0-3.2 mm . . . .
. . . . 28. *lineolatus* Mannerheim
- Segment 1 of mid tarsi not strongly excised ventrally
(Figs 208-211). Elytral apex more narrowly rounded or
truncate. Elytral lateral border rather narrow posteriorly
(Fig. 216). Penis with a broadly or narrowly rounded apex
(Figs 200-203) 20
20(19) Front tarsal claws of subequal shape and size (Fig. 211).
Right paramere with a ventral fringe of long hairs which
does not reach the apex of the paramere (Fig. 203).
Length: 2.2-2.8 mm . . . . 33. *heydeni* Wehncke
- Inner front tarsal claw shorter, wider and more strongly
curved than the outer claw (Figs 208-210). Right paramere
with a ventral fringe of long hairs which reaches the apex
of the paramere (Figs 200-202) 21
21(20) Body rather wide; widest point usually just behind prono-
tum, sides converging rather strongly behind (Fig. 175).
Pronotal basal plicae short and curved (Fig. 183). Transverse row of punctures on abdominal sternite VI not wi-

dely interrupted in the middle (Fig. 217). Yellow colour of body often with a reddish tinge. The interrupted and confluent dark lines usually give the elytra a pronounced

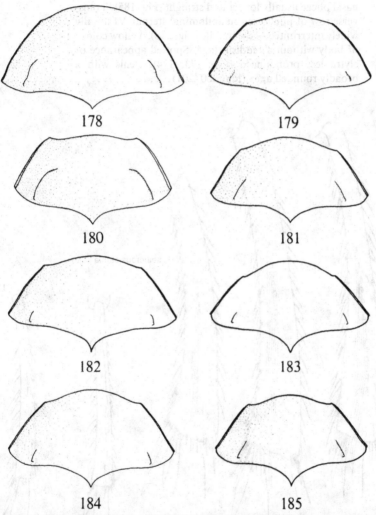

Figs 178-185. Pronotum of *Haliplus* spp. – 178: *H. confinis* Stph.; 179: *H. varius* Nicol.; 180: *H. lineatocollis* (Marsh.); 181: *H. apicalis* Thoms.; 182: *H. fluviatilis* Aubé; 183: *H. ruficollis* (De Geer); 184: *H. immaculatus* Gerh.; 185: *H. wehnckei* Gerh.

spotted appearance (Fig. 175). Penis tapering to a narrowly rounded apex (Fig. 202). Length: 2.5-3.0 mm . . . .
. . . . 32. *ruficollis* (De Geer)
–   Body usually narrower, widest near the middle and with sides more evenly rounded (Figs 173, 174). Pronotal basal plicae usually longer and straight (Fig. 185). Transverse row of punctures on abdominal sternite VI usually widely interrupted in the middle (Fig. 218). Yellow colour of body without a reddish tinge. Spotted appearance of elytra less pronounced (Figs 173, 174). Penis with a broadly rounded apex (Figs 200, 201) 22

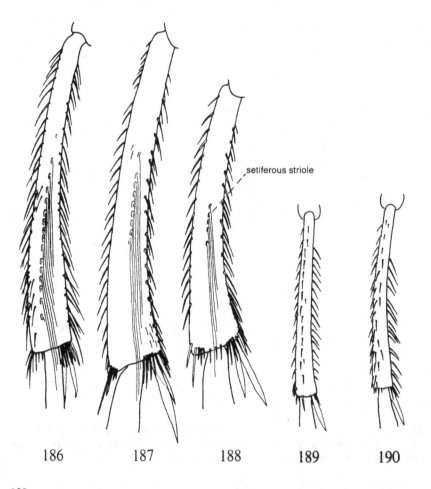

setiferous striole

186   187   188   189   190

22(21) Ventral side of right paramere only weakly dilated pre-
        apically (Fig. 201). Length: 2.5-3.3 mm .... 31. *wehnckei* Gerhardt
  –   Ventral side of right paramere more strongly dilated pre-
        apically (Fig. 200). Length: 2.8-3.0 mm .... 30. *sibiricus* Motschulsky
23(16) Elytral apex broadly rounded or truncate. Elytral lateral
        border rather wide posteriorly (Fig. 215). Elytra usually
        (but not always) with micro-punctuation covering the
        entire surface. Length: 2.0-3.2 mm .... 28. *lineolatus* Mannerheim
  –   Elytral apex more narrowly rounded or truncate. Elytral
        lateral border rather narrow posteriorly (Fig. 216). Elytra
        with or without micro-punctuation 24
24(23) Segment 5 of front tarsi usually with 3 ventral spines, and
        segment 5 of mid tarsi usually with 4 ventral spines (Fig.
        213); the spines are rather long, and the proximal ones are
        more closely set than the distal ones. Body fairly short,
        with rather evenly rounded sides (Fig. 172). Elytral dark
        lines usually strongly interrupted, giving the elytra a pale,
        spotted appearance (Fig. 172). Length: 2.4-2.8 mm....................
        .... 29. *interjectus* Lindberg

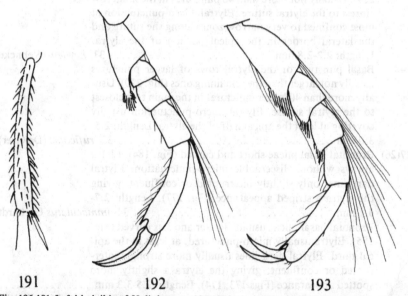

191   192   193

Figs 186-191. Left hind tibia of *Haliplus* spp.; 186-188 in dorsal view, 189-191 in dorsolateral view.
– 186: *H. laminatus* (Schall.); 187: *H. fulvus* (F.); 188: *H. variegatus* Sturm; 189: *H. lineatocollis*
(Marsh.); 190: *H. confinis* Stph.; 191: *H. ruficollis* (De Geer).
Figs 192, 193. Mid tarsus of ♂ of 192: *Haliplus laminatus* (Schall.) and 193: *H. fulvus* (F.).

–   Segment 5 of front and mid tarsi usually with more spines; these are rather evenly dispersed and usually shorter (Fig. 214). Shape of body and elytral dark pattern variable (Figs 173-177).. 25

25(24) Body rather short; usually widest just behind pronotum, converging rather strongly behind (Figs 175, 176). Pronotal basal plicae short (Fig. 183). The interrupted and confluent dark lines usually give the elytra a pronounced spotted appearance (Figs 175, 176). Transverse row of punctures on abdominal sternite VI not widely interrupted in the middle (Fig. 217).. 26

–   Body narrower, usually widest near the middle, with more evenly rounded sides (Figs 173, 174, 177). Length of pronotal basal plicae variable (Figs 184, 185). Spotted appearance of elytra less pronounced (Figs 173, 174, 177). Transverse row of punctures on abdominal sternite VI usually widely interrupted in the middle (Fig. 218)................. 27

26(25) Some of the elytral rows of larger punctures often with the basal puncture much larger than the remaining ones (Fig. 219). Usually not more than 40 punctures in the main row closest to the elytral suture. Elytral micro-punctuation at most confined to very narrow zones along the suture and the lateral border in the apical portion of the elytra. Length: 2.2-2.8 mm .... 33. *heydeni* Wehncke

–   Basal puncture of the elytral rows of larger punctures usually not larger than the remaining ones (Fig. 220). Usually more than 40 larger punctures in the main row closest to the elytral suture. Elytral micro-punctuation usually covering at least the apical half of the elytra. Length: 2.5-3.0 mm .... 32. *ruficollis* (De Geer)

27(26) Pronotal basal plicae short and curved (Fig. 184). Elytra always without discernable micro-punctuation. Elytral dark lines only slightly interrupted or confluent, giving the elytra a striped appearance (Fig. 177). Length: 2.7-3.1 mm................................. 34. *immaculatus* Gerhardt

–   Pronotal basal plicae usually longer and less curved (Fig. 185). Elytra usually micro-punctured, at least in the apical third. Elytral dark lines usually more strongly interrupted or confluent, giving the elytra a slightly more spotted appearance (Figs 173, 174). Length: 2.5-3.3 mm ....
.... 30. *sibiricus* Motschulsky
or 31. *wehnckei* Gerhardt

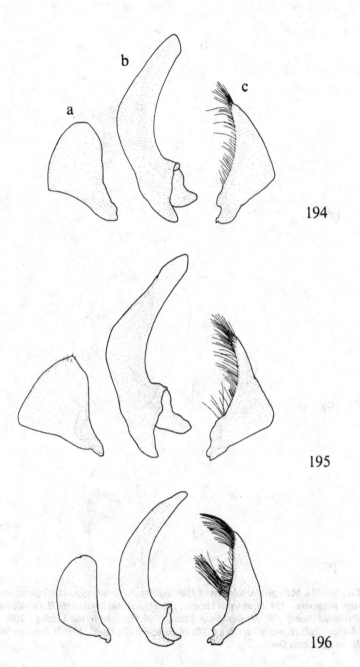

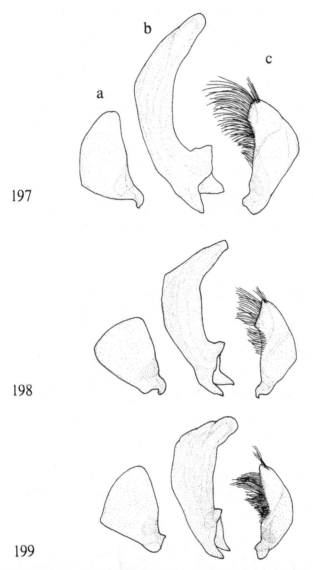

b

c

a

197

198

199

Figs 194-204. Male genital sclerites of *Haliplus* (Sg. *Haliplus*) spp.; a: left paramere; b: penis; c: right paramere. – 194: *H. apicalis* Thoms.; 195: *H. furcatus* Seidl.; 196: *H. fulvicollis* Er.; 197: *H. fluviatilis* Aubé; 198: *H. lineolatus* Mann.; 199: *H. interjectus* Lindbg.; 200: *H. sibiricus* Motsch.; 201: *H. wehnckei* Gerh.; 202: *H. ruficollis* (De Geer); 203: *H. heydeni* Wehncke; 204: *H. immaculatus* Gerh.

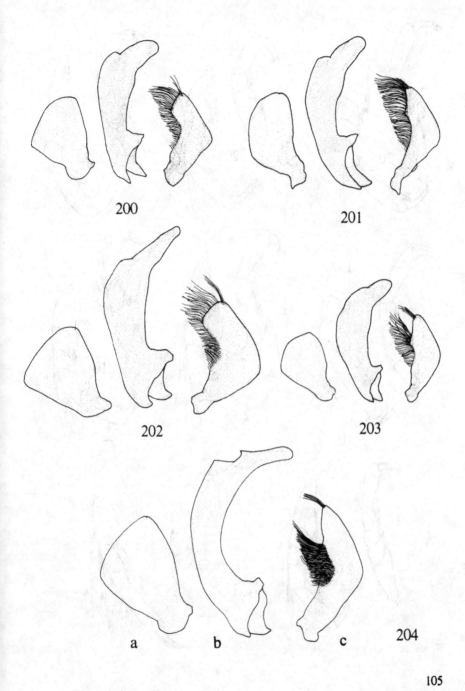

200

201

202

203

a   b   c   204

105

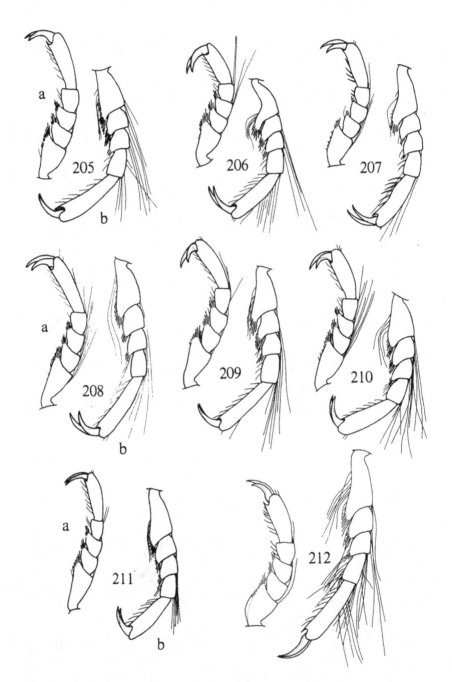

205   206   207

208   209   210

211   212

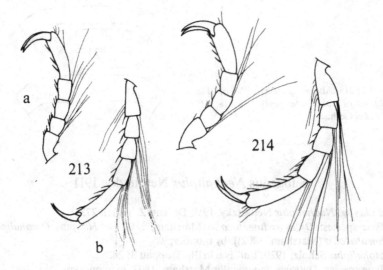

Figs 205-214. Front tarsus (a) and mid tarsus (b) of *Haliplus* (Sg. *Haliplus*) spp. – 205: *H. fluviatilis* Aubé ♂, 206: *H. lineolatus* Mann. ♂, 207: *H. interjectus* Lindbg. ♂, 208: *H. sibiricus* Motsch. ♂, 209: *H. wehnckei* Gerh. ♂, 210: *H. ruficollis* (De Geer) ♂, 211: *H. heydeni* Wehncke ♂, 212: *H. immaculatus* Gerh. ♂, 213: *H. interjectus* Lindbg. ♀; 214: *H. immaculatus* Gerh. ♀.

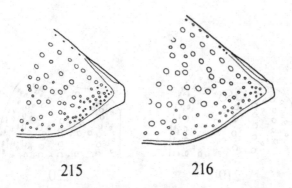

Figs 215, 216. Posterior part of left elytron of 215: *Haliplus lineolatus* Mann. and 216: *H. immaculatus* Gerh.

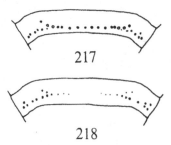

217

Figs 217, 218. Abdominal sternite VI of 217: *Haliplus ruficollis* (De Geer) and 218: *H. wehnckei* Gerh.

218

Subgenus *Neohaliplus* Netolitzky, 1911

Haliplus sg. *Neohaliplus* Netolitzky, 1911, Dt. ent. Z. (1911): 273.
Type species: *Dytiscus lineatocollis* Marsham, 1802 [= *Haliplus (Neohaliplus) lineatocollis* (Marsham, 1802)], by monotypy.
Protohaliplus Scholz, 1929, Ent. NachrBl., Troppau 3: 28.
Type species: *Dytiscus lineatocollis* Marsham, 1802, by monotypy.

Small or medium-sized species. Pronotum and elytra without discernable micro-punctuation. Penultimate segment of labial palps very wide in the European species, with one or two denticles along the inner edge. Pronotum with very long, curved basal ·plicae in the European species (Fig. 180). Posterior angles of pronotum slightly pro-truding laterally. Hind tibia with two rather regular single rows of spines along the out-er edge, without a setiferous striole on the upper face (Fig. 189).

Two species from the western Palearctic region and three species from Australia have been assigned to this subgenus.

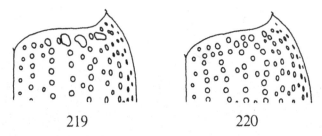

219   220

Figs 219, 220. Basal part of right elytron of 219: *Haliplus heydeni* Wehncke and 220: *H. ruficollis* (De Geer).

108

16. *Haliplus (Neohaliplus) lineatocollis* (Marsham, 1802)
Figs 124, 131, 136, 157, 180, 189, 221-225.

Dytiscus lineatocollis Marsham, 1802, Ent. Brit. 1: 429.
Haliplus lineaticollis, sensu auctt.; incorrect spelling.

2.3-3.3 mm. Rather elongate, sides of body almost parallel in the middle. Dorsal side somewhat flattened, yellowish, often with head and median parts of pronotum darkened; elytra usually with a number of dark blotches, rarely with continuous dark lines following the rows of larger, black punctures (Fig. 157). Ventral side largely reddish, legs yellowish. Prosternal apophysis not bordered laterally, but deeply sulcate torwards apex. Metasternal apophysis almost even, with a depressed, slightly oblique row of punctures on each side of the middle. Male: segments 1-3 of front and mid tarsi slightly dilated; front claws simple, of subequal length; penis broadly rounded apically; left paramere with several long hairs at apex; right paramere with a fringe of long hairs along the ventral edge and at apex; genital sclerites, vide Figs 221-223. Female: gonocoxosternites with rather short and wide struts; tergal halves IX rather short and wide; gonocoxae short, triangular; genital sclerites, vide Figs 224, 225.

Distribution. Recorded from all Danish districts, but not common in the northwest. – In the south of Sweden: Sk., Hall., Öl., Gtl., Vg. – Not in Norway or East Fennoscandia. – Most of Europe, north to Britain, Fennoscandia, Poland and central parts of European USSR; the Canaries, N. Africa, Ethiopia, Asia Minor, Israel, Lebanon, Syria, Yemen, Saudi Arabia, and the Caucasus.

Biology. Mainly in base-rich ponds and streams. The larva, and perhaps also the

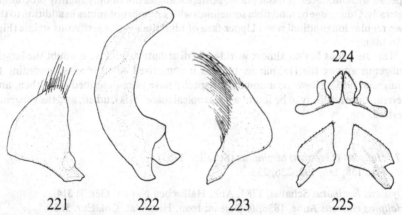

Figs 221-223. Male genital sclerites of *Haliplus lineatocollis* (Marsh.). – 221: left paramere; 222: penis; 223: right paramere.
Figs 224, 225. Female genital sclerites of *Haliplus lineatocollis* (Marsh.). – 224: gonocoxae and tergal halves IX; 225: gonocoxosternites.

adult, feeds on filamentous algae. Eggs are deposited in the spring or summer, and after a short period the young larvae appear. In southern Europe the life-cycle may be completed the same year, but in Fennoscandia the larva usually hibernates, and pupates the following spring. The young adults appear after a few weeks. The pupa and the larval instars were first described by Bertrand (1923); further details on the larval instars are provided by Seeger (1971a). In Fennoscandia the larva often hibernates out of the water, while the adult usually hibernates in the water. In southern Europe the species has been recorded from altitudes up to 2200 m (Franciscolo, 1979). It has been observed flying on many occasions.

Subgenus *Liaphlus* Guignot, 1928

Haliplus sg. *Liaphlus* Guignot, 1928, Annls Soc. ent. Fr. 97: 138.
  Type species: *Dytiscus fulvus* Fabricius, 1801 [= *Haliplus (Liaphlus) fulvus* (Fabricius, 1801)], by subsequent designation (Guignot, 1930).
Haliplus sg. *Hoplites* Kinel, 1929, Polsk. Pismo entomol. 8: 219; preocc., *nec* Neumayr, 1875 (Cephalopoda), (see note under *H. laminatus*).
  Type species: *Dytiscus laminatus* Schaller, 1783 [= *Haliplus (Liaphlus) laminatus* (Schaller, 1783)], by monotypy.

Medium- or large sized species. Pronotum and elytra usually without discernable micro-punctuation; in Fennoscandia only females of *laminatus* have micro-punctured elytra. Penultimate segment of labial palps rather narrow and unmodified, except in *laminatus* which has this segment widened as in sg. *Haliplus* s. str. Pronotum in our species without traces of basal plicae, posterior angles not or only slightly protruding laterally. Outer edge of hind tibia sometimes with a number of spines in addition to the two regular longitudinal rows. Upper face of hind tibia with a setiferous striole (Figs 186-188).
  This subgenus has an almost world-wide distribution, and is no doubt the largest subgenus among the Haliplidae. In 1955 it comprised 40-50 species, according to Guignot (1955). However, a number of species have been described since then, and several more are likely to be found when tropical material is studied, and the subgenus revised.

17. *Haliplus (Liaphlus) laminatus* (Schaller, 1783)
  Figs 158, 186, 192, 226, 233.

Dytiscus laminatus Schaller, 1783, Abh. Hallischen Naturf. Ges. 1: 314.
Haliplus cinereus Aubé, 1836, *in* Dejean: Icon. Hist. Nat. Col. Eur. 5: 30.
Haliplus affinis, sensu auctt.; misident., *nec* Stephens, 1829.

2.5-3.3 mm. Oblong, body widest close to the middle; outline markedly constricted between pronotum and elytra. Dorsal side normally yellowish with dark punctures and

110

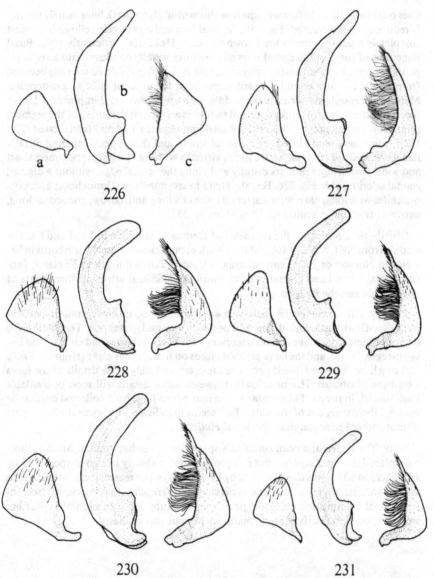

Figs 226-231. Male genital sclerites of *Haliplus* (Sg. *Liaphlus*) spp.; a: left paramere; b: penis; c: right paramere. – 226: *H. laminatus* (Schall.); 227: *H. flavicollis* Sturm; 228: *H. mucronatus* Stph.; 229: *H. fulvus* (F.), specimen from Denmark; 230: same species, specimen from northern Sweden; 231: *H. variegatus* Sturm.

lines on the elytra, and often with the head brownish; elytral dark lines usually strong-ly reduced and interrupted (Fig. 158). Ventral side and legs largely yellowish. Almost completely black specimens are known to occur. Head not particularly wide. Basal punctures of some of the elytral rows of punctures usually confluent into large com-pound punctures. Prosternal apophysis sulcate and weakly bordered laterally between front coxae; its sides extend forward as plicae to the anterior edge of prosternum. Metasternal apophysis sulcate in the middle, with a wide transverse impression. Upper face of hind tibia with a setiferous striole which is about half the length of the segment (Fig. 186). Male: segments 1-3 of front tarsi and segment 1 of mid tarsi dilated (Fig. 192); front claws unmodified, of subequal length and shape; penis tapering distally into a rounded apex; left paramere short, virtually without hairs; right paramere short and wide, with fringes of hairs distally and along the ventral edge, without a digitus; genital sclerites, vide Fig. 226. Female: elytra micro-punctured throughout; gonocox-osternites with long, narrow struts; tergal halves IX long and narrow; gonocoxae long, narrowly triangular; genital sclerites, vide Fig. 233.

Distribution. Mainly in the southeast of Denmark; common in LFM and SZ, few records from NEJ, NWZ, NEZ and B. – Sweden: only known from a few records in Sk. – Not in Norway or East Fennoscandia. – Central Europe, north to England, Fen-noscandia, Poland and central parts of European USSR, south to northern parts of Spain, Italy and the Balkans.

Biology. This species prefers base-rich water without too much vegetation, particu-larly slow-flowing streams, drains and ponds in clay and gravel pits. The adult feeds on both filamentous algae and invertebrates such as oligochaetes and chironomid lar-vae (Seeger, 1971b), and the larva probably feeds on filamentous algae (Holmen, 1981). Not much has been published upon the life-cycle, and only a few details of the larva have been illustrated (Holmen, 1981). However, more details will soon be available (van Vondel, in press). Third instar larvae and adults have been collected during the winter, hibernating out of the water. This seems to indicate a life-cycle similar to that of most other Fennoscandian species of *Haliplus*.

Note. The systematic position of this species is somewhat unclear. Most authors have placed it in sg. *Liaphlus,* but it is probably more closely related to species of sg. *Haliplus* s. str. This was already noted by Kinel (1929) who created a separate subgenus for this species; however, the name of his subgenus *(Hoplites)* is preoccupied (see: In-ternational Commission on Zoological Nomenclature, 1955: opinion 353). Further work is needed to clarify the systematic position of this species.

Figs 232-237. Female genital sclerites of *Haliplus* (Sg. *Liaphlus*) spp.; a: gonocoxae; b: tergal halves IX; c: gonocoxosternites. – 232: *H. flavicollis* Sturm; 233: *H. laminatus* (Schall.); 234: *H. mucronatus* Stph.; 235: *H. fulvus* (F.), specimen from Denmark; 236: same species, specimen from northern Sweden; 237: *H. variegatus* Sturm.

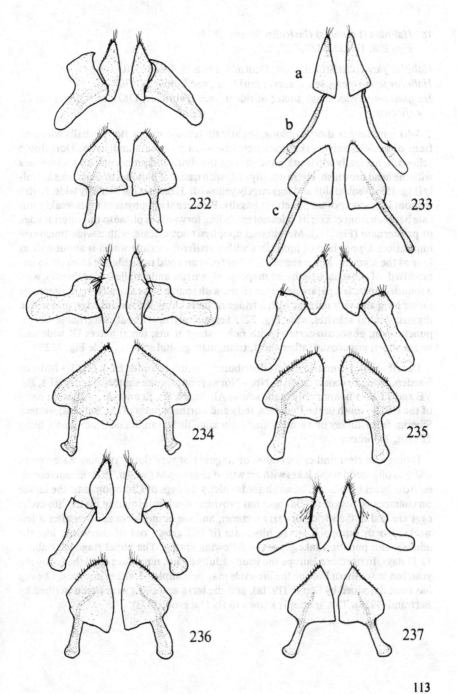

232 233

234 235

236 237

18. *Haliplus (Liaphlus) flavicollis* Sturm, 1834
Figs 108, 118, 151, 159, 227, 232.

Haliplus flavicollis Sturm, 1834, Deutschl. Faun. 5, Ins. 8: 150.
Haliplus ferrugineus, sensu auctt.; misident., *nec* Linnaeus, 1767.
Haliplus impressus, sensu auctt.; misident., *nec* (Fabricius, 1787); see note under *H. ruficollis.*

3.4-4.1 mm. Rather shortly oblong, slightly flattened dorsally. Body usually widest in front of middle, only weakly constricted between pronotum and elytra. Dorsal side yellowish, normally with dark punctures on the elytra and pronotum, and sometimes with the head brownish; elytra usually only with traces of blotches or dark lines apically (Fig. 159). Ventral side and legs largely yellowish. Head not particularly wide. Elytra without large compound punctures basally. Prosternal apophysis at most weakly sulcate between front coxae; its sides not extending forwards as plicae to the anterior edge of prosternum (Fig. 151). Metasternal apophysis not sulcate, with a wide transverse impression. Upper face of hind tibia with a setiferous striole which is about half as long as the segment. Male: segments 1-3 of front and mid tarsi dilated; front claws unmodified, of subequal length and shape; penis narrow and parallel-sided distally, with a rounded apex; left paramere rather short, with sparse hairs apically; right paramere rather long and very narrow, with a fringe of hairs along the ventral edge, and with a digitus; genital sclerites, vide Fig. 227. Female: elytra without discernable micropunctuation; gonocoxosternites with rather short struts; tergal halves IX wide and rather short; gonocoxae rather short, triangular; genital sclerites, vide Fig. 232.

Distribution. Denmark: widely distributed, but not recorded from NWJ. – Most of Sweden, along the coast north to Nb. – Norway: only in the southeast: AK, HEs, Bø, VE and TEi. – Finland: only in the south: Al, Ab, N, Ka, St and Sa. – Adjacent parts of the USSR, south to the Pyrenees, Italy and northern parts of the Balkans; western Siberia, from Turkey onwards to Transcaucasia, Turkestan, Kazakhstan, and China; ?Egypt, ?Morocco.

Biology. In clear and clean bodies of stagnant or very slowly running water, especially in oligoproductive lakes with growth of characeans (Seeger, 1971c); even recorded from brackish water. The adult feeds chiefly on eggs of Chironomidae, the larvae on characeans. Seeger (1971a, b, c) has provided much information on the life-cycle: eggs are laid in the spring or early summer, and the young larva appears after a few weeks. North European larvae hibernate (if full grown out of the water, like the adults), and pupation take place the following spring. The pupal stage lasts about 12-13 days. In northern Europe the young adults do not reproduce until the following year, but in southern Europe the life-cycle may be completed within one year. The egg has been described by Seeger (1971a), and the larva and pupa were first described by Bertrand (1928). This species is known to fly (Jackson, 1973).

114

Haliplus (Liaphlus) mucronatus Stephens, 1828
  Figs 109-111, 119, 160, 228, 234.

Haliplus mucronatus Stephens, 1829, Ill. Brit. Ent., Mandib. 2: 40.
Haliplus parallelus Babington, 1836, Trans. ent. Soc. London 1: 178.

3.7-4.3 mm. Resembling the preceding species, but differing in the following combination of characters: body oblong and rather wide behind, almost parallel-sided in the middle. Elytra without dark lines and at most with weak traces of blotches along the suture (Fig. 160). Head very wide; the distance between the eyes is at least twice the width of an eye. Elytra often with large compound punctures basally. Sides of prosternal apophysis extending forwards as plicae to the anterior edge of prosternum. Metasternal apophysis weakly sulcate in the middle, with a shallow, wide transverse impression. Upper face of hind tibia with a setiferous striole of more than half the length of the segment. Male: penis wide, of almost same width throughout most of its length, apex broadly rounded; right paramere rather wide; genital sclerites, vide Fig. 228. Female: genital sclerites, vide Fig. 234.

Distribution. Not recorded from Denmark and Fennoscandia. – Southern and central Europe, north to England and German F. R. and D. R., east to Czechoslovakia, Austria, the Balkans, Turkey and the USSR (along the Black Sea); N. Africa along the Mediterranean Sea.

Biology. Mainly found in base-rich stagnant water, both fresh and saline; often in clay and gravel pits. The larva was first described by Bertrand (1928) (as *H. guttatus*); it probably feeds on characeans.

19. *Haliplus (Liaphlus) fulvus* (Fabricius, 1801)
  Figs 117, 137, 150, 161, 162, 187, 193, 229, 230, 235, 236; pl. 1: 4.

Dytiscus fulvus Fabricius, 1801, Syst. Eleuth. 1: 271.
Dytiscus interpunctatus Marsham, 1802, Ent. Brit. 1: 429.
Haliplus Lapponum Thomson, 1855, Svenska Vet. Akad. Handl. (1854): 184.
Haliplus Lapponum var. *niger* Seidlitz, 1887, Verh. naturf. Ver., Brünn 25: 31.
Haliplus fulvus var. *unicolor* Munster, 1922, Videnskapsselsk. Skr. 1 (9): 14.
Haliplus fulvus var. *Sparre-schneideri* Munster, 1922, Videnskapsselsk. Skr. 1 (9): 15.
?*Haliplus (Liaphlus) salinarius* Wallis, 1933, Trans. R. Can. Inst. 21: 56; see note.
Haliplus ferrugineus, sensu auctt.; misident., *nec* (Linnaeus, 1767).

3.5-4.6 mm. Resembling the two preceding species, but differing in the following combination of characters: body oblong, widest near the middle, sides rather evenly rounded. Dorsum normally largely yellowish, usually with dark punctures and a number of blotches on the elytra; often also with posterior part of head and margins and basal punctures on pronotum dark; elytral dark blotches often very prominent, but not, or only to a small extent, confluent with the dark suture (Figs 161, 162). Ventral side and legs usually largely yellowish or reddish. The coloration of this species varies a lot, and almost unicolourous dark or yellowish specimens are known to occur. Head not particularly wide. Elytra without large compound punctures basally. Sides of prosternal apophysis extending forwards as plicae to the anterior edge of prosternum. Metasternal apophysis with a pit-shaped depression in the middle, otherwise almost even (Fig. 150). Upper face of hind tibia with a setiferous striole of about one third the length of the segment, consisting of about 10 punctures (Fig. 187). Male: penis weakly narrowed distally, with a rounded apex; right paramere rather wide; genital sclerites, vide Figs 229, 230. Female: genital sclerites, vide Figs 235, 236.

Distribution. All of Fennoscandia and Denmark, though not recorded from a few districts. – Most of Europe, including Iceland and the Faroes; Turkey, Siberia; western N. America.

Biology. Chiefly in clear, clean bodies of stagnant, and especially in the north, in slowly running water; particularly found in oligoproductive, slightly acid lakes with growth of characeans (Seeger, 1971c), but even recorded from brackish water. The larva probably feeds on characeans. The life-cycle is much as described for *flavicollis*, except for the fact that north European adults seem to emerge from the pupae very early in the spring (Seeger, 1971c). The larva was first described by Schiødte (1864). The species is known to fly (Jackson, 1973).

Notes. N. American specimens, probably representing *H. (Liaphlus) salinarius* Wallis, have been examined; they are undoubtedly conspecific with *H. (Liaphlus) fulvus* (Fabr.). However, no type-material of *H. (Liaphlus) salinarius* Wallis has been studied.

Specimens from higher elevations and from northern parts of the range, and particularly forund in running waters (Huldén, 1983b), differ from typical specimens (particularly found in stagnant habitats) as follows: shape of body narrower, especially along the basal portion of the elytra (Fig. 162); punctures generally more coarse and more deeply impressed; dark colouration of the body more strongly developed (Fig. 162); anterior edge of pronotum more strongly protruding in the middle; pronotum with a rather deep transverse basal depression; micro-punctuation on hind tibia more coarse and dense; male genital sclerites narrower (Fig. 230). This variation seems to be clinal, at least in northern Europe, and the subdivision into separate subspecies or species, as proposed by some authors, is avoided.

20. *Haliplus (Liaphlus) variegatus* Sturm, 1834
    Figs 134, 163, 188, 231, 237.

Haliplus variegatus Sturm, 1834, Deutschl. Faun. 5, Ins. 8: 157.
Haliplus subnubilus Babington, 1836, Trans. ent. Soc. London 1: 177.
Haliplus rubicundus Babington, 1836, Trans. ent. Soc. London 1: 178.

2.5-3.5 mm. Resembling the three preceding species, but differing as follows: body usually rather shortly oval, widest slightly in front of the middle. Dorsal side largely yellowish or reddish, usually with dark punctures and a number of blotches on the elytra; often also with posterior part of the head, and margins and basal punctures on pronotum dark; elytral blotches normally very prominent, confluent with each other and with the dark suture (Fig. 163). Ventral side and legs largely yellowish or reddish. Unicolorous yellowish specimens are known. Head not particularly wide. Elytra without large compound punctures basally. Sides of the prosternal apophysis extending forward as plicae to the anterior edge of prosternum. Metasternal apophysis with a pit-shaped depression in the middle, otherwise almost even. Upper face of hind tibia with a setiferous striole of about one fifth the length of the segment, consisting of about 5 punctures (Fig. 188). Male: genital sclerites (Fig. 231) strongly resembling those of *fulvus*, though the distal part of penis is slightly narrower. Female: genital sclerites, vide Fig. 237.

Distribution. Recorded from all Danish districts, but more common in the southeast. – Sweden: locally abundant in the south, mainly near the coast: Sk., Bl., Öl., Gtl., Ög., Vg., Boh., Upl. – Norway: only in the southeast, very rare: Ø, VE. – Finland: only in the southwest, mainly near the coast, locally common: Al, Ab, N. – Not in adjacent parts of the USSR. – Central and southern Europe, north to Britain, Fennoscandia and central and southern parts of European USSR; Cyprus, Turkey, Israel, Syria, Iran, Iraq, Afghanistan, N. Africa (see note).

Biology. Usually met with in smaller stagnant bodies of water, particularly in habitats with clear water and dense submerged vegetation; often found near the coast, but also in old peat-cuttings. The morphology of the larva suggests that it feeds on characeans. Not much is known about its life-cycle. Larvae have only been obtained during the summer, and this seems to indicate that only adults hibernate. Hibernating adults have been collected from ice-covered habitats during the winter. The larva was first described by Schiødte (1864).

Note. The distribution towards the east and south is somewhat uncertain. A number of no doubt separate species, closely related to *variegatus*, occur in the eastern Mediterranean area and in the Middle East; these have been considered conspecific with *variegatus* by many authors. All records of *variegatus*, sensu auctt. have been included in the distribution given above, but the distribution of the true *variegatus* is probably more limited.

Subgenus *Haliplidius* Guignot, 1928

Haliplus sg. *Haliplidius* Guignot, 1928, Annls Soc. ent. Fr. 97: 138.
   Type species: *Dytiscus obliquus* Fabricius, 1787 [= *Haliplus (Haliplidius) obliquus* (Fabricius, 1787)], by subsequent designation (Guignot, 1930).
Haliplus sg. *Haliplus* s. str., sensu Guignot, 1939, Bull. Soc. ent. Fr. 44: 176.

Small or medium-sized species. Most of body covered by strong micro-punctuation. Penultimate segment of labial palps somewhat widened, with a denticle or angle along

117

the inner edge. Pronotal basal plicae weakly impressed or absent. Posterior angles of pronotum hardly protruding laterally. Outer edge of hind tibia either with irregularly set spines (Fig. 190) or with spines arranged in longitudinal rows as in Figs 189, 191. Hind tibia without a setiferous striole on the upper face.

Three species from the Palearctic region and one from southernmost S. America have been assigned to this subgenus.

21. *Haliplus (Haliplidius) confinis* Stephens, 1829
   Figs 112, 120, 135, 149, 164, 178, 190, 238, 241.

Haliplus confinis Stephens, 1829, Ill. Brit. Ent., Mandib. 2: 41.
Haliplus lineatus Aubé, 1836, *in* Dejean: Icon. Hist. Nat. Col. Eur. 5: 21.
Haliplus confinis var. *pallens* Fowler, 1887, Col. Brit. Isl. 1: 153.
Haliplus confinis f. *Hellieseni* Munster, 1922, Videnskapsselsk. Skr. 1 (9): 11.
Haliplus confinis var. *halberti* Bullock, 1928, Entomologist's mon. Mag. 64: 103.

3.0-3.8 mm. Oblong, body widest closely behind pronotum. Dorsal side yellowish, usually with dark lines on the elytra, and sometimes also posterior part of head and anterior and posterior margins of pronotum brownish; elytral dark lines usually only interrupted to a small extent, and often somewhat confluent (Fig. 164). Ventral side and legs largely yellowish or reddish. Larger punctures rather shallow, on the elytra arranged in very irregular rows (Fig. 112). Prosternal apophysis bordered laterally (Fig. 149). Metasternal apophysis almost even, with a shallow depression on each side of the middle. Male: segments 1-3 of front and mid tarsi dilated; the inner front claw slightly wider and shorter than the outer; aedeagus very large; penis narrow in the distal part, with an ax-shaped apex; left paramere hardly with any hairs; right paramere with a fringe of short hairs along the ventral edge; genital sclerites, vide Fig. 238. Female: gonocoxosternites with long, narrow struts; tergal halves IX rather long and narrow; gonocoxae long, narrowly triangular; genital sclerites, vide Fig. 241.

Distribution. All of Denmark. – Sweden: widely distributed, though not recorded from a few districts. – Norway: seems to be rare in the southwest; recorded from Ø, AK, HEn, VAy, STy, STi, Nsy, Nsi, Nnø, Nnv and TRi. – Most of East Fennoscandia, though not recorded from a few districts. – Europe, north to Britain and Fennoscandia, south to the Pyrenees, northern Italy and the Balkans; northern and central USSR, eastwards to Siberia, Transcaucasia, Turkey.

Biology. Mainly in clear stagnant bodies of water such as lakes and larger ponds; but also in the backwaters of slow-flowing streams and in brackish water. Often found in depths of more than one m., and often among characeans, on which the larva is known to feed. Eggs are deposited in the spring or early summer, and after a few weeks the young larvae appear. In Fennoscandia the larva hibernates out of the water and it pupates the following spring. The young adults appear after a few weeks. The larva was first described by Schiødte (1864) (as *H. ruficollis*), and the pupa by Bertrand (1928).

118

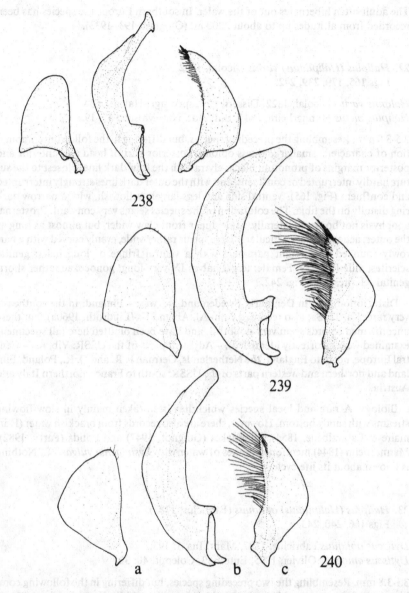

238

239

a   b   c   240

Figs 238-240. Male genital sclerites of *Haliplus* (Sg. *Haliplidius*) spp.; a: left paramere; b: penis; c: right paramere: – 238: *H. confinis* Stph.; 239: *H. varius* Nicol.; 240: *H. obliquus* (F.). 238 is in half size compared to 239 and 240.

The adult often hibernates out of the water. In southern Europe the species has been recorded from altitudes up to about 2200 m. (Guignot, 1931-1933).

22. *Haliplus (Haliplidius) varius* Nicolai, 1822
    Figs 165, 179, 239, 242.

Haliplus varius Nicolai, 1822, Dissert. Col. spec. agri Halensis: 34.
Haliplus pictus Mannerheim, 1844, Bull. Soc. Nat. Moscou 17: 192.

2.5-3.0 mm. Resembling the preceding species, but differing in the following combination of characters: smaller and less oblong. Posterior part of head and anterior and posterior margins of pronotum black; elytra with the four dark lines closest to the suture hardly interrupted or confluent, and with the outer dark lines strongly interrupted and confluent (Fig. 165). Ventral side and legs largely yellowish, with a narrow dark ring distally on the tibia. The coloration of the species seems very constant. Prosternal apophysis not bordered laterally. Male: inner front claw wider, but almost as long as the outer; aedeagus not particularly large; penis rather wide, evenly curved with a narrowly rounded apex; right paramere with a ventral fringe of long hairs; genital sclerites, vide Fig. 239. Female: tergal halves IX very long; gonocoxae rather short; genital sclerites, vide Fig. 242.

Distribution. Not in Denmark, Sweden and Norway. – Finland: in the southeast, very rare: Ka, Ta, Sa; also recorded from Al, Ab and Ks (Lindroth, 1960a), but these unconfirmed records seem very doubtful, and have been omitted here (all specimens examined were incorrectly identified). – Adjacent parts of the USSR: Vib, Kr. – Central Europe, north to England, the Netherlands, German F. R. and D. R., Poland, Finland and northern and western parts of the USSR; south to France, northern Italy and Austria.

Biology. A rare and local species which has been taken mainly in slow-flowing streams with sandy bottom. However, there are also records from brackish water (Fairmaire & Laboulbène, 1854), dune-lakes (Guignot, 1947) and ponds (Parry, 1982); Mannerheim (1844) mentioned bodies of water with *Lysimachia vulgaris* L. Nothing is known about its life-cycle.

23. *Haliplus (Haliplidius) obliquus* (Fabricius, 1787)
    Figs 166, 240, 243.

Dytiscus obliquus Fabricius, 1787, Mant. Ins. 1: 193.
Dytiscus amoenus Olivier, 1795, Entom. 3, Coleopt. 40: 32.

3.1-3.8 mm. Resembling the two preceding species, but differing in the following combination of characters: oblong; body widest near the middle. Posterior part of head and posterior and anterior margins of pronotum sometimes brownish; elytral dark lines normally present, but usually strongly interrupted, and often confluent (Fig.

120

166). Ventral side and legs largely yellowish; tibiae without a distal black ring. Prosternal apophysis not bordered laterally. Metasternal apophysis weakly impressed in the middle. Male: the inner front claw very slightly shorter and narrower than the outer; aedeagus not particularly large; penis rather narrow in the distal part, with a rounded apex; right paramere with a fringe of long hairs along the ventral edge; genital sclerites, vide Fig. 240. Female: tergal halves IX and gonocoxae long; genital sclerites, vide Fig. 243.

Distribùtion. Denmark: widely distributed, but rare in the west and north; not in WJ. – Sweden: mainly in the southeast, north to Upl., but even recorded from Jmt. – Norway: only one record: Nsi, Susendalen, 1974 (Dolmen & Koksvik, 1976). – Finland: only in the southwest, near the coast: Al, Ab, N, St. – Not in adjacent parts of

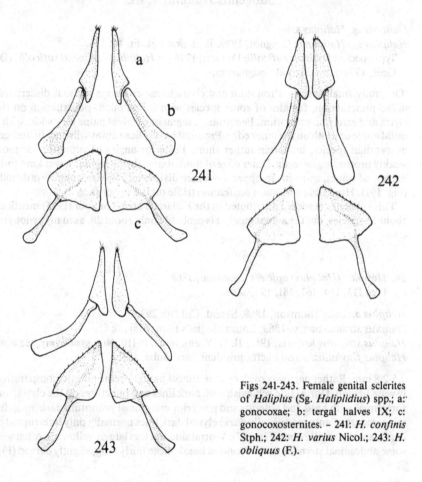

Figs 241-243. Female genital sclerites of *Haliplus* (Sg. *Haliplidius*) spp.; a:' gonocoxae; b: tergal halves IX; c: gonocoxosternites. – 241: *H. confinis* Stph.; 242: *H. varius* Nicol.; 243: *H. obliquus* (F.).

the USSR. – Most of Europe, north to Britain, Fennoscandia and parts of European USSR, south to Spain, Italy and the Balkans; Morocco, Turkey, Iran, Transcaucasia, ?Siberia.

Biology. Prefers clear, base-rich, stagnant bodies of water; often found in lakes and gravel-pits, and it even occurs in brackish water. The species often lives in depths of more than one m., and usually among characeans, on which the larva feeds. According to Seeger (1971c) the life-cycle is largely as described for *flavicollis*. The larva and pupa were first described by Bertrand (1928). This species has been collected at elevations up to 2300 m. in the Alps (Guignot, 1931-1933).

Subgenus *Haliplus* s. str.

Haliplus sg. *Haliplus* s. str.
Haliplus sg. *Haliplinus* Guignot, 1939, Bull. Soc. ent. Fr. 44: 176.
   Type species: *Dytiscus ruficollis* De Geer, 1774 [= *Haliplus (Haliplus) ruficollis* (De Geer, 1774)], by original designation.

Generally small species. Pronotum and elytra in males always without discernable micro-punctuation, females of some species with a fine micro-punctuation on the elytra and rarely on pronotum. Penultimate segment of labial palps very wide, with a distal angle or tooth on the inner edge. Pronotal basal plicae always distinct in the Fennoscandian species, but often rather short. Posterior angles of pronotum at most ·weakly protruding laterally. Outer edge of hind tibia with one regular lower longitudinal row of spines, and with an upper row which divides into two irregular rows distally (Fig. 191). Hind tibia without a setiferous striole on the upper face.

This subgenus is widely distributed in the Holarctic area. Guignot (1955) mentions about 55 species, but the actual number is probably only about 28, as many synonyms exist.

24. *Haliplus (Haliplus) apicalis* Thomson, 1868
   Figs 113, 154, 167, 181, 194, 244.

Haliplus apicalis Thomson, 1868, Skand. Col. 10: 293.
Haliplus striatus Sharp, 1869, Entomologist's mon. Mag. 6: 81.
Haliplus strigatus Roberts, 1913, Jl. N. Y. ent. Soc. 21: 110. **New synonymy**, see note.
Haliplus fluviatilis, sensu auctt.; misident., *nec* Aubé, 1836.

2.5-2.8 mm. Rather narrowly oval, outline almost parallel in the middle, comparatively wide behind. Dorsal side yellowish with dark lines and punctures on the elytra, and sometimes with head and anterior and posterior margins of pronotum darkened; distal segments of antennae usually dark; elytral dark lines normally only interrupted or confluent to a small extent (Fig. 167). Ventral side and legs largely yellowish, often with some abdominal sternites dark. Pronotal basal plicae fairly long, slightly curved (Fig.

181). Elytra with the presutural punctures much finer than the punctures of the main row closest to the suture. Prosternal apophysis with a weak median furrow basally which may branch into two, weakly bordering the sides of the apophysis in its anterior part (Fig. 154). Metasternal apophysis usually with a longitudinal, slightly oblique depression on each side of the middle, otherwise almost even. Male: segments 1-3 of front and mid tarsi dilated; inner front tarsal claw slightly shorter, wider, and a little more strongly curved than the outer; penis rather evenly rounded, with a truncate apex; left paramere with hardly any hairs; right paramere triangular, narrow distally, with a fringe of hairs along most of the ventral edge; genital sclerites, vide Fig. 194. Female: elytra totally covered by micro-punctuation (Fig. 113); struts of gonocoxosternites, tergal halves IX and gonocoxae rather long and narrow; tergal halves IX a little shorter than gonocoxae; genital sclerites, vide Fig. 244.

Distribution. Common in Denmark along the coast; not in B. – Sweden: along the southern coasts, north to Boh. and Gtl. – Norway: a few records along the southern coast: Ø, AK, Bø, VAy. – Not in East Fennoscandia. – Britain, Belgium, the Netherlands, German F. R. and D. R., Hungary; northern parts of European and Siberian USSR, Mongolia, western N. America.

Biology. Mainly in stagnant or slowly running water with high salinity. It chiefly occurs near the coast (most European records) and in areas with steppe. The larva, and to some extent also the adult, feeds on filamentous algae (Seeger, 1971b). Eggs are deposited in the spring or summer, and the larva pupates later the same summer. The adults hibernate out of the water. The larva was first described by Bertrand (1942).

Note on synonymy. Syntypes of *Haliplus strigatus* Roberts, 1913, deposited in the American Museum of Natural History, New York, and syntypes of *Haliplus apicalis* Thomson, 1869, deposited in the Zoological Museum, Lund, have been examined, and proved to be conspecific.

25. Haliplus (Haliplus) furcatus Seidlitz, 1887
Figs 153, 168, 195, 245.

Haliplus furcatus Seidlitz, 1887, Verh. naturf. Ver., Brünn 25: 33.
Haliplus fluviatilis var. *Mannerheimii* Seidlitz, 1887, Verh. naturf. Ver., Brünn 25: 33.

2.2-2.8 mm. Resembling the preceding species, but differing as follows: sides of the elytra more evenly rounded. Primary colour of body often reddish. All antennal segments usually yellowish; elytral dark lines normally both strongly interrupted and confluent (Fig. 168). Abdominal sternites at most weakly darkened. Pronotal basal plicae sometimes rather short. Elytral presutural punctures at least basally of almost same size as the punctures of the main row closest to the suture. Prosternal apophysis with a rather deep median furrow basally which branches into two, bordering the sides of the apophysis through most of its length (Fig. 153). Metasternal apophysis with a longitudinal, often pit-shaped depression on each side of the middle, otherwise almost even. Male: penis very long, distal part slightly incised dorsally, apex obliquely trun-

123

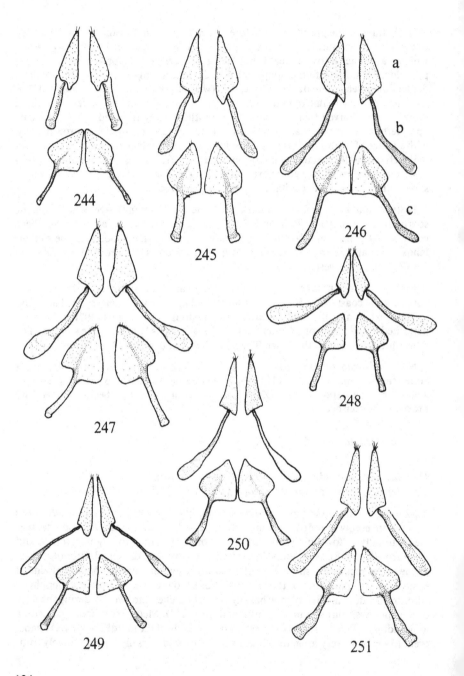

244

245

246
a
b
c

247

248

249

250

251

cate; genital sclerites, vide Fig. 195. Female: elytra sometimes with a micro-punctuation which may cover most of their surface; tergal halves IX about as long as the gonocoxae; genital sclerites, vide Fig. 245.

Distribution. Rare and local in southern Denmark: WJ, F, LFM, SZ, NEZ, B. - In the southeast of Sweden, mainly near the coast; rather common in Öl., but also recorded from Sk., Gtl., Ög. and Upl. - Not in Norway. - Finland: only one record: Al, Finström, Bjärström, 1 ♂, 30.VI.1943 (Håkan Lindberg). - Not in adjacent parts of the USSR. - England, Belgium, the Netherlands, German F. R. and D. R., Poland, Czechoslovakia, Austria, northern Italy; parts of the USSR from Europe to Siberia.

Biology. Especially in sun-exposed marshes with temporary water. Often found in habitats near the sea or in areas with steppe, though it seems to avoid very high salinities. The larva probably feeds on filamentous algae. Females with large eggs in the ovaries have been obtained in May, second instar larvae in June, and teneral adults in April. The larva has not yet been described. Adults hibernate out of the water.

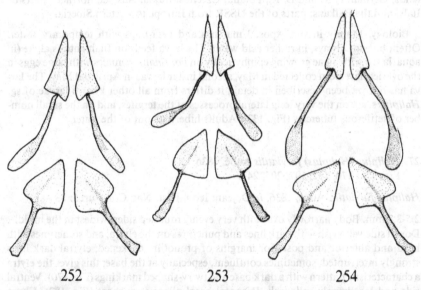

252   253   254

Figs 244-254. Female genital sclerites of Haliplus (Sg. Haliplus) spp.; a: gonocoxae; b: tergal halves IX; c: gonocoxosternites. - 244: H. apicalis Thoms.; 245: H. furcatus Seidl.; 246: H. fulvicollis Er.; 247: H. fluviatilis Aubé; 248: H. lineolatus Mann.; 249: H. interjectus Lindbg.; 250: H. sibiricus Motsch.; 251: H. wehnckei Gerh.; 252: H. ruficollis (De Geer); 253: H. heydeni Wehncke; 254: H. immaculatus Gerh.

125

26. *Haliplus (Haliplus) fulvicollis* Erichson, 1837
   Figs 133, 152, 169, 196, 246.

Haliplus fulvicollis Erichson, 1837, Käf. Mark Brandenb. 1: 186.
Haliplus jakowlewi Semenov, 1897, Trudy russk. ent. Obshch. 31: 545.

2.6-3.0 mm. Resembling the preceding species, but differing as follows: body usually more widely oval, with evenly rounded sides of elytra. Elytral dark lines very strongly interrupted and confluent, normally forming a number of brown blotches (Fig. 169). Prosternal apophysis not furrowed or bordered (Fig. 152). Male: penis not particularly long, with the sides rather evenly rounded, distally tapering into the narrowly rounded apex; ventral edge of right paramere with a short fringe of long hairs in the middle and another at the apex; genital sclerites, vide Fig. 196. Female: elytra without discernable micro-punctuation; tergal halves IX longer than gonocoxae; genital sclerites, vide Fig. 246.

Distribution. Southern parts of Denmark, rare: SJ, EJ, WJ, LFM, SZ and NEZ. – Sweden: rare, but found in most districts north to Upl. – Norway: only one locality: Ø, Kirkeøy, Hvaler, 1925-26 (Munster, 1927). – Finland: known only from the southwest: Al (many records), St. – Adjacent parts of the USSR: Vib, Kr. – Belgium, the Netherlands, German F. R. and D. R., Poland, Czechoslovakia, Austria, northern parts of Italy, and the Balkans; parts of the USSR from Europe to western Siberia.

Biology. Mainly in sun-exposed marshes and peat-bogs with temporary water. Often, but not always, in rather acid water. The larva feeds on filamentous algae (in aquaria fed with algae growing epiphytically on *Fontinalis*. Females with large eggs in the ovaries have been collected in May, second instar larvae in April and May. The larva has not yet been described in detail; it differs from all other known larvae of sg. *Haliplus* s. str. on the very long lateral processes of the tergites, and on the small number of setiferous tubercles (Fig. 133). Adults hibernate out of the water.

27. *Haliplus (Haliplus) fluviatilis* Aubé, 1836
   Figs 156, 170, 182, 197, 205, 247.

Haliplus fluviatilis Aubé, 1836, *in* Dejean: Icon. Hist. Nat. Col. Eur. 5: 33.

2.5-3.2 mm. Body narrowly oval with very evenly rounded sides, widest in the middle. Dorsal side yellowish with dark lines and punctures on the elytra, and sometimes with head and anterior and posterior margins of pronotum darkened; elytral dark lines strongly interrupted, sometimes confluent, especially at the base; this gives the elytra a characteristic pattern with a dark base and two v-shaped markings (Fig. 170). Ventral side and legs largely yellowish. Pronotal basal plicae very short (Fig. 182). Elytra sometimes with a number of basal punctures confluent into larger compound punctures. Elytral apex a little more narrowly truncate or rounded than in *lineolatus*. Elytral lateral border rather narrow posteriorly. Prosternal apophysis not, or only very weakly furrowed medially. Metasternal apophysis sometimes slightly furrowed at apex, and at most with a small depression in the middle, otherwise almost even (Fig. 156). The row of punctures across abdominal sternite VI usually not interrupted in the middle. Male: segments 1-3 of front and mid tarsi dilated, segments 1 not particularly modified; inner front tarsal claw slightly shorter, slightly more curved and slightly wider than the outer (Fig. 205); penis with rather evenly curved sides and a broadly rounded apex; left paramere with hardly any hairs; right paramere triangular, ventral edge somewhat dilated preapically, and with a fringe of long hairs which reaches the apex; genital sclerites, vide Fig. 197. Female: elytra micro-punctured for at least posterior two thirds; struts of gonocoxosternites, tergal halves IX and gonocoxae rather long and narrow; tergal halves IX slightly longer than gonocoxae; genital sclerites, vide Fig. 247.

Distribution. Most of Denmark, not in NWJ and B. – In the southeast of Sweden: Sk., Gtl., Ög., Sdm. and Upl.; previous records from Öl. (Lindroth, 1960a) have been omitted, as the specimens examined proved to belong to other species (mainly *heydeni*). – Not in Norway. – In the southeast of Finland: Ab, N, Ta, Sa and Oa. – Adjacent parts of the USSR: Vib, Kr. – Most of Europe, north to Britain, Fennoscandia and northern and central parts of European USSR; Turkey, the Urals, Transcaucasia, Kirgiziya, Afghanistan, Siberia.

Biology. Chiefly found among vegetation in larger streams, more rarely in larger ponds and lakes with clear water and sandy or stony bottom. The larva probably feeds on filamentous algae. In Fennoscandia both larvae and adults hibernate, at least the latter out of the water. It is not certain whether the larva described as *fluviatilis* by Gernet (1868) actually belongs to this species; Bertrand (1950) later provided a description of the larva.

28. *Haliplus (Haliplus) lineolatus* Mannerheim, 1844
Figs 171, 198, 206, 215, 248.

Haliplus lineolatus Mannerheim, 1844, Bull. Soc. Nat. Moscou 17: 190; see note.
Haliplus Schaumi Solsky, 1867, Trudy russk. ent. Obshch. 5: 39: **New synonymy**, see note.
Haliplus transversus Thomson, 1870, Opusc. ent. 2: 124; see note.
Haliplus nomax Balfour-Browne, 1911, Entomologist's mon. Mag. 47: 153; see note.
Haliplus browneanus Sharp, 1913, Entomologist's mon. Mag. 49: 75; emend. of *brownei;* see note.
Haliplus nomax var. *laevigatus* Falkenström, 1936, Ent. Tidskr. 57: 252.
Haliplus fluviatilis, sensu auctt.; misident., *nec* Aubé, 1836.

2.0-3.2 mm. Resembling the preceding and the six following species, but differing as follows: body rather widely oval, particularly behind, with sides less evenly rounded. Elytral dark lines usually not very strongly interrupted or confluent (Fig. 171), rarely obsolete. Pronotal basal plicae often rather long, almost straight. Elytra normally without basal compound punctures. Elytral apex widely rounded or truncate. Elytral lateral border wide posteriorly (Fig. 215). Prosternal apophysis with a shallow median furrow, at least anteriorly. Metasternal apophysis furrowed anteriorly, with a large, deep depression in the middle. The row of punctures across abdominal sternite VI sometimes interrupted in the middle. Male: segment 1 of mid tarsi very strongly incised ventrally (Fig. 206); front tarsi and claws almost as in *fluviatilis;* penis bent in an angle distally, with the sides tapering into a rather narrowly truncate apex; distal dilation of right paramere not shaped as in *fluviatilis;* genital sclerites, vide Fig. 198. Female: most of elytra usually micro-punctured, but specimens without micro-punctuation occur; tergal halves IX much longer than gonocoxae; genital sclerites, vide Fig. 248.

Distribution. All of Denmark, except NWZ; locally common. - Widely distributed in Sweden, but not recorded from a few districts. - Norway: rather few and scattered records: AK, HEs, Os, On, Bø, HOy, NTi and Fi. - All of Finland and adjacent parts of the USSR. - Britain, Belgium, the Netherlands, German F. R. and D. R., Poland, northern parts of European USSR; Siberia, Mongolia.

Biology. Mainly found in clear, oxygen-rich lakes, and in the backwaters of larger streams; also taken in brackish water. It is a characteristic species of weakly euproductive stagnant waters (Seeger, 1971c). The adult chiefly feeds on hydrozoans, the larva on filamentous algae (Seeger, 1971a, b). The life-cycle in northern Europe is largely as described for *flavicollis* (Seeger, 1971c). The larva described as *fluviatilis* by Gernet (1868) may belong to this species; Seeger (1971a) later described the larva of *lineolatus* in details. *H. lineolatus* is known to fly (Jackson, 1973).

Notes on synonymy. Various authors have considered specimens of both *lineolatus* and *wehnckei* to be types of *Haliplus lineolatus* Mannerheim, 1844. No lectotype has yet been designated, but it appears less likely that the single specimen of *wehnckei,* deposited in the Zoologische Staatssammlung, München, was among Mannerheim's syntypes. The majority of specimens which might qualify as syntypes, deposited in the Naturhistoriska Riksmuseet, Stockholm, and in the Zoological Museum, University of Copenhagen, belong to *lineolatus* in the present sense. Furthermore, the stability of the nomenclarure would best be supported by the designation of a specimen of *lineolatus* as lectotype. This problem has been discussed by Balfour-Browne (1936a, 1940) and Falkenström (1939).

Two specimens, deposited in the Zoological Institute, Academy of Sciences, Leningrad, are here designated as lectotype and paralectotype of *Haliplus schaumi* Solsky, 1867. - Lectotype, ♂ USSR: Kosloff, 3275-3. - Paralectotype: USSR: Kosloff, 3275-2 (1 ♀). This species is in my opinion conspecific with *lineolatus.*

One specimen of *lineolatus* and one of *heydeni* have both been considered to be the type-specimen of *H. transversus* Thomson, 1870. The former specimen, previously deposited in Berlin, is now missing from Thomson's collections, and no lectotype has been designated. The view of Balfour-Browne (1940a) regarding the synonymy is followed here, as it best supports the stability of the nomenclature.

30 specimens from F. Balfour-Browne's collection, deposited in the Royal Scottish

Museum, Edinburgh, are here designated as lectotype and paralectotypes of *Haliplus nomax* Balfour-Browne, 1911. – Lectotype, ♂: Northern Ireland: Co. Down, Carrickmannan Lough, 18.V.1911. – Paralectotypes: same data as lectotype (22 specim.); Nothern Ireland: Co. Armagh, Maghery, 25.X.1908 (1 specim.); Northern Ireland: Co. Antrim, Canal at Toome, 16.IV.1910 (1 specim.); Eire: Co. Carlow, 12.IV.1911 (1 specim.); Northern Ireland: Co. Down, 6.V.1911 (3 specim.); Northern Ireland: Co. Antrim, Shaw's Bridge, Lagan Canal, 16.V.1911 (1 specim.).

Two specimens from Sharp's collection, deposited in the British Museum (Natural History), are here designated as lectotype and paralectotype of *Haliplus browneanus* Sharp, 1913. – Lectotype ♂ and paralectotype ♀: England: 1167, Ouse, 7.IX.1868.

29. *Haliplus (Haliplus) interjectus* Lindberg, 1937
Figs 172, 199, 207, 213, 249.

Haliplus interjectus Lindberg, 1937, Acta Soc. Fauna Flora fenn. 60: 487; see note.
Haliplus (Haliplinus) transversus, sensu Zaitsev, 1953 (in part); misident., *nec* Thomson, 1870; see note.

2.4-2.8 mm. Resembling the preceding and the five following species, differing in the following combination of characters: body shortly oval, widest near the middle, with evenly rounded sides. Elytral dark lines normally strongly interrupted and reduced, giving the elytra a pale, spotted appearance (Fig. 172). Pronotal basal plicae rather short, slightly curved. Elytra usually without larger compound punctures basally. Elytral apex narrowly rounded or truncate. Elytral lateral border narrow posteriorly. Segment 5 of front and mid tarsi (Figs 207, 213) differ from all other members of the subgenus: ventral side not straight in males; ventral spines not evenly dispersed and erect (closer set and more erect basally); very few ventral spines (male, front: 4-5; male, mid: 5-6; female, front: 3-4; female, mid: 4-5). The row of punctures across abdominal sternite VI is usually interrupted in the middle. Male: segment 1 of front tarsi wide basally, with a ventral row of short spines, without a ventral longitudinal ridge (Fig. 207); the inner front claw distinctly shorter, more curved and wider than the outer; segment 1 of mid tarsus markedly incised ventrally; penis bent in an angle distally, with a short distal part and a widely rounded apex; ventral edge of right paramere hardly dilated distally, with a fringe of hairs which does not reach the apex, and an apical tuft of hairs; genital sclerites, vide Fig. 199. Female: elytra sometimes with micropunctuation, at most occupying the posterior two thirds; tergal halves IX somewhat longer than gonocoxae; genital sclerites, vide Fig. 249.

Distribution. Not in Denmark, Sweden and Norway. – Finland: probably not uncommon in the south and east; recorded from Ab, N, Ka, Ta, Sa, Ks. – Adjacent parts of the USSR: Vib, Kr, Lr. – Northern parts of European USSR, Altai, Yakutian ASSR, Kamtchatka; probably widely distributed in the USSR.

Biology. Records include ponds (Tkačenko, 1929), clay-pits with *Carex*-hummocks (Rutanen, *in litt.*)., lakes, a shallow vegetation-rich cove of the Baltic Sea (Lindberg,

1937), streams (Silfverberg, *in litt.;* Tkačenko, l.c.), and a hot spring (water +35°C, air –10°C; Zimmermann, 1925). Nothing is otherwise known about the biology.

Notes. Specimens of *interjectus* have often been mistaken for other species, probably due to the very brief original description (Lindberg, 1937). Material from Kamtchatka was assigned to the N. American *H. robertsi* Zimm. by Zimmermann (1925), and specimens from the USSR were included in *H. lineolatus* Mann. (as *transversus* Thoms.) by Zaitsev (1953). Finnish specimens have usually been identified as *H. wehnckei* Gerh.

Two specimens, deposited in the Zoological Museum, University of Helsinki, are here designated as lectotype and paralectotype of *Haliplus interjectus* Lindberg, 1937. – Lectotype, ♂ and paralectotype, ♂: USSR: Viborg, 24.VIII.1934, P. H. Lindberg. One ♀ with the same data, deposited in the Zoological Institute, Academy of Sciences, Leningrad, is here designated as paralectotype.

30. *Haliplus (Haliplus) sibiricus* Motschulsky, 1860
Figs 173, 200, 208, 250.

Haliplus sibiricus Motschulsky, 1860, *in* Schrenck: Reisen und Forsch. im Amurlande 2: 99; see note.
Haliplus Sahlbergi Falkenström, 1939, Ark. Zool. 32A (6): 32.
Haliplus Lindbergi Falkenström, 1939, Ark. Zool. 32A (6): 38.
Haliplus transversus, sensu auctt.; misident., *nec* Thomson, 1870.
Haliplus lineolatus, sensu auctt.; misident., *nec* Mannerheim, 1844.
Haliplus ruficollis, sensu auctt.; misident., *nec* (De Geer, 1774).

2.8-3.0 mm. A very variable species, resembling the two preceding and the four following species; differing in the following combination of characters: body rather narrowly oval, widest near the middle, sides rather evenly rounded. Elytral dark lines somewhat interrupted, not often confluent; the lines are strongly abbreviated in the Fennoscandian specimens, giving the elytra a pale, slightly spotted or striped appearance (Fig. 173). Pronotal basal plicae often rather long and straight. Elytra often with larger compound punctures basally, these normally not as large as in *heydeni.* Apex of elytra rather narrowly truncate or rounded. Elytral lateral border narrow posteriorly. The row of punctures across abdominal sternite VI usually widely interrupted in the middle. Male: segment 1 of front and mid tarsi not particularly modified (Fig. 208); the inner front claw distinctly shorter, more curved and wider than the outer; penis not very evenly curved, with an outstanding extension or "hood" on the dorsal side, and a broadly rounded apex; ventral edge of right paramere with a fairly large angular dilation distally, and with a fringe of hairs which reaches the apex; genital sclerites, vide Fig. 200. Female: elytra sometimes with micro-punctuation, occupying at most the posterior two thirds; tergal halves IX somewhat longer than the gonocoxae; genital sclerites, vide Fig. 250.

Distribution. Not in Denmark. – Sweden: a few records in the north, mainly from

130

the Torneå river system: Nb. and T. Lpm. – Norway: only one record from Fi, Alten, vi.1924 (Munster). – In the north of Finland: recorded from ObN in the Torneå river system, and from Li. – Adjacent parts of the USSR: Vib, Kr, Lr. – Northern parts of the USSR from Europe to Siberia; Altai, Mongolia.

Biology. Almost all records are from running water, otherwise nothing is known.

Notes on synonymy. Motschulsky (1853) first published the name *Haliplus sibiricus* as a nomen nudum, giving only the distribution of the species: "Siberia in general". In the valid original description of the species (Motschulsky, 1860) the distribution was given as: "Collected by M. Schrenck in the waters of the Amur, but also common in the rest of southern Siberia".

Zaitsev (1915) considered a syntype labelled "100. Schrenck's R." to be the type of the species, and assigned it correctly to *H. ruficollis* (De G.). However, he later considered *sibiricus* to be a senior synonym of *H. sahlbergi* Falk. instead, and provided a description of it, also perfectly fitting Falkenström's species (Zaitsev, 1953).

I prefer to follow the latter view, as this species better fits the original description of *sibiricus,* and as the name is most frequently used in this sense.

Five other syntypes from Tobolsk were assigned to *H. fluviatilis* Aubé by Zaitsev (1915). They actually belong to *H. sibiricus* Motsch. (2 specim.), *H. ruficollis* (De G.) (2 specim.) and *H. lineolatus* Mann. (1 specim.).

One of these specimens, which are all deposited in the Zoological Institute, Academy of Sciences, Leningrad, is here designated as lectotype of *Haliplus sibiricus* Motschulsky, 1860. – Lectotype, ♂: USSR: G. Tobolsk, Bileika.

The *H. sibiricus,* sensu Zimmermann (1924) is a different species, known only from eastern Siberia (Holmen, *in prep.*). Unfortunately Zaitsev (1953) adapted Zimmermann's key for the parts of his key concerning *sibiricus.*

31. *Haliplus (Haliplus) wehnckei* Gerhardt, 1877
    Figs 116, 174, 185, 201, 209, 218, 251.

Haliplus borealis Gerhardt, 1877, Z. Ent. 6: 36; preocc., *nec* LeConte, 1850.
Haliplus Wehnckei Gerhardt, 1877, Dt. ent. Z. 21: 448; replacement name.
Haliplus immaculatus, sensu auctt., misident., *nec* Gerhardt, 1877.
Haliplus lineolatus, sensu auctt.; misident., *nec* Mannerheim, 1844; see note under *H. lineolatus.*
Haliplus transversus, sensu auctt.; misident., *nec* Thomson, 1870.

2.5-3.3 mm. Very strongly resembling *sibiricus* (see note), and only separated from this species by the shape of the right paramere in the male: the distal dilation of the ventral edge is somewhat weaker in *wehnckei,* though it may be a little more strongly developed than shown in Fig. 201, particularly in northern specimens. Dark lines of elytra usually more well developed than in *sibiricus,* especially in southern specimens (Fig. 174). Almost completely black specimens are known. Pronotal basal plicae rather long and straight (Fig. 185). The row of punctures across abdominal sternite VI usually

131

widely interrupted in the middle (Fig. 218). Male: front and mid legs, vide Fig. 209; genital sclerites, vide Fig. 201. Female: genital sclerites, vide Fig. 251.

Distribution. All of Denmark. – Sweden, Norway and East Fennoscandia: widely distributed, but absent from many districts. – Most of Europe, though rare in the south, north to Britain, Fennoscandia and northern and central parts of European USSR, south to the Pyrenees and northern parts of the Balkans and Italy; western Siberia, Mongolia.

Biology. Mainly found in running water, especially in smaller, not too swift streams; also in lakes and larger ponds with clear water. Beier (1929) provides much information about the biology: the adult feeds on both filamentous algae and invertebrates (e.g. larvae of Chironomidae and Oligochaeta), and the larva on filamentous algae. Eggs are deposited throughout the summer, mainly inside the stems of submerged vascular plants, and the young larvae appear after a few weeks. According to Beier (l. c.), the species has at least two generations per year in Austria. In southern Fennoscandia the larvae may be found throughout most of the year. The majority of larvae and adults hibernate out of the water. The larva was described by Beier (l.c.). The species is known to fly (Jackson, 1973). Adults are capable of stridulation (Beier, l. c.), using the laterosternal edges as a plectrum for a band of fine spines on the elytra (Fig. 116).

Note. Brinck (1944) proposed that *wehnckei* (as *lineolatus*) and *sibiricus* (as *sahlbergi*) should be treated as subspecies of one species. It is very likely that they are actually conspecific, but whether they should be considered separate subspecies cannot be proven, until more material has shown that they are not sympatric, and that they are not merely parts of a clinally varying species. For the time being they are treated as separate species, following the majority of recent authors.

32. *Haliplus (Haliplus) ruficollis* (De Geer, 1774)
Figs 107, 114, 115, 155, 175, 183, 191, 202, 210, 217, 220, 252; pl. 1: 3.

Dytiscus ruficollis De Geer, 1774, Mem. Hist. Ins. 4: 404.
Dytiscus impressus Fabricius, 1787, Mant. Ins. 1: 193; preocc., *nec* Müller, 1776 (Dytiscidae); see note.
Dytiscus marginepunctatus Panzer, 1794, Faun. Germ. 1 (14): t. 10.
Haliplus affinis Stephens, 1829, Ill. Brit. Ent., Mandib. 2: 42; see note.

2.5-3.0 mm. Resembling the four preceding and the two following species; may be separated from these on the following combination of characters: body rather shortly oval, usually widest in front of middle, sides narrowed rather strongly behind. Primary colour reddish or yellowish. Dark lines of elytra widely interrupted and often confluent, giving the elytra a spotted appearance (Fig. 175). Basal plicae of pronotum rather short and curved (Fig. 183). Elytra normally without basal compound punctures (Fig. 220). Elytral rows of large punctures usually with more punctures than in *heydeni* (the main row closest to the suture with more than 40 punctures). Apex of

132

elytra narrowly truncate or rounded. Elytral lateral border narrow posteriorly. The row of punctures across abdominal sternite VI rarely interrupted in the middle (Fig. 217). Male: segment 1 of front and mid tarsi not particularly modified (Fig. 210); the inner front claw distinctly shorter, wider and much more strongly curved than the outer; penis rather long, bent in an angle distally, tapering into a narrowly rounded apex, and with an extension or "hood" on the dorsal side; ventral edge of right paramere with a large angular dilation distally, and with a fringe of hairs which reaches the apex; genital sclerites, vide Fig. 202. Female: elytra usually with micro-punctuation which rarely occupies more than the posterior two thirds (Fig. 115); tergal halves IX somewhat longer than the gonocoxae; genital sclerites, vide Fig. 252.

Distribution. All of Denmark, very common. – Common in the south of Sweden, mainly near the coast in the north. – In most of Norway, excepting the northernmost parts and at higher elevations. – Finland: common in most of the country, but not reaching the northernmost parts. – Adjacent parts of the USSR: Vib, Kr, Lr. – Most of Europe, north to Britain, Fennoscandia and parts of European USSR, south to the Mediterranean Sea; Turkey, Iran, Transcaucasia, Kazakhstan, western Siberia.

Biology. Chiefly found in smaller bodies of stagnant water with rich vegetation, but also in lakes and streams and in brackish water. It is a characteristic species of poly-productive stagnant waters (Seeger, 1971c), and it does not seem very sensitive to pollution. The larva, and to a large extent also the adult, feeds on filamentous algae (Seeger, 1971b). Eggs may be deposited through most of the summer (Seeger, 1971b), inside emptied cells of submerged plants. The young larvae appear after a few weeks; some larvae probably pupate later the same summer, but many hibernate to pupate the following spring. The pupal stage lasts a couple of weeks, and the young adults probably do not reproduce until the following year. Hibernation of larvae and adults normally takes place out of the water. The larva and pupa were first described by Bertrand (1928); the larva described by Schiødte (1864) as *ruficollis* actually belongs to *confinis*. *H. ruficollis* is known to fly (Jackson, 1973).

Notes on synonymy. The specimens of *Dytiscus impressus* in Fabricius' collection, deposited in the Zoological Museum, University of Copenhagen, presently comprises the following species: *Haliplus flavicollis* Sturm, 1834 (1 specim.), *H. immaculatus* Gerhardt, 1877 (2 specim.) and *H. ruficollis* (De Geer, 1774) (1 specim.). According to Erichson (1837), specimen(s) of *H. lineatocollis* (Marsham, 1802) were previously also present. To preserve the stability of the nomenclature, the specimen of *ruficollis* is here designated as lectotype of *Dytiscus impressus* Fabricius, 1787. – Lectotype, ♀: (only my lectotype label). – None of the above mentioned specimens presently carry any other labels, and they all fit the original description of *impressus*. Erichson (1837) argued that the specimen of *flavicollis* would best qualify as type, but the size of *impressus* stated by Fabricius (1801) under *Dytiscus fulvus* does not fit *flavicollis* very well.

A specimen from Stephens' collection, deposited in the British Museum (Natural History), is here designated as lectotype of *Haliplus affinis* Stephens, 1829. – Lectotype, ♀: (only my lectotype label). – None of the seven specimens of *affinis* in the

133

Stephens collection carried any other labels. As they all fit the original description of *affinis,* a specimen of *H. ruficollis* (De Geer, 1774) was selected as lectotype, to preserve the stability of the nomenclature. It was possible to examine three of the remaining six specimens, and they belong to *H. wehnckei* Gerhardt, 1877 (2 specim.) and *H. fluviatilis* Aubé, 1836 (1 specim.).

33. *Haliplus (Haliplus) heydeni* Wehncke, 1875
Figs 176, 203, 211, 219, 253.

Haliplus Heydeni Wehncke, 1875, Dt. ent. Z. 19: 122.
Haliplus foveostriatus Thomson, 1884, Opusc. Ent. 10: 1030.
Haliplus transversus, sensu auctt.; misident., *nec* Thomson, 1870; see note under *H. lineolatus.*

2.2-2.8 mm. Resembling the six preceding and the following species (especially *ruficollis*); recognized on the following combination of characters: body rather shortly oval, usually widest in front of the middle, sides narrowed rather strongly behind. Dark lines of elytra widely interrupted and often confluent, giving the elytra a spotted appearance (Fig. 176). Basal plicae of pronotum rather short and curved. Elytra usually with large compound punctures basally (Fig. 219). Elytral rows of large punctures usually with fewer punctures than in *ruficollis* (the main row closest to the suture usually with less than 40 punctures). Apex of elytra narrowly rounded or truncate. Elytral lateral border narrow posteriorly. Metasternal apophysis usually with a large depression in the middle, but sometimes almost even as in *fluviatilis.* The row of punctures across abdominal sternite VI rarely interrupted in the middle. Male: segment 1 of front and mid tarsi not particularly modified; front claws of subequal length and shape (Fig. 211); penis rather short, not evenly rounded, with a short, apically rounded distal part, and with a dorsal extension or "hood"; ventral edge of right paramere hardly dilated distally, with a fringe of hairs which does not reach the apex, and an apical tuft of hairs; genital sclerites, vide Fig. 203. Female: elytra at most micropunctured posteriorly in narrow zones near the lateral border and the suture; tergal halves IX somewhat longer than gonocoxae; genital sclerites, vide Fig. 253.

Distribution. All of Denmark. - Sweden: widely distributed in the southern and middle parts, north to Ång. - Only in the south and southeast of Norway: Ø, AK, Os, Bø, TEy, AAy, VAy. - Finland: mainly in the south, north to about latitude 83°. - Adjacent parts of the USSR: Vib, Kr. - Most of Europe, north to England, Fennoscandia and northern and central parts of European USSR, south to northern parts of Spain, Italy and the Balkans; Turkey, the Caucasus, Kazakhstan, Turkestan, western Siberia.

Biology. Mainly found in small bodies of stagnant or slowly running water with rich vegetation; known both from very acid water and from brackish water. It is a characteristic species of poly-productive stagnant waters (Seeger, 1971c). The larva, and to a large extent also the adult, feeds on filamentous algae (Seeger, 1971b). The life-cycle seems largely to be as in *ruficollis* (Seeger, 1971a, b). The larva was described by Bertrand (1942).

134

34. *Haliplus (Haliplus) immaculatus* Gerhardt, 1877
Figs 132, 177, 184, 204, 212, 214, 216, 254.

Haliplus immaculatus Gerhardt, 1877, Z. Ent. 6: 36.
Haliplus fluviatilis, sensu auctt.; misident., *nec* Aubé, 1836.
Haliplus striatus, sensu auctt.; misident., *nec* Sharp, 1869.

2.7-3.1 mm. Resembling the six preceding species, but recognized on the following combination of characters: body rather narrowly oval, widest near the middle, sides rather strongly narrowed behind. Elytral dark lines normally only weakly interrupted or confluent, giving the elytra a striped appearance (Fig. 177); the dark lines are rarely strongly reduced. Pronotum with basal plicae short and often curved (Fig. 184). Elytra occasionally with slightly larger compound punctures basally. Apex of elytra rather narrowly rounded or truncate. Elytral lateral border rather narrow posteriorly (Fig. 216). The row of punctures across abdominal sternite VI usually widely interrupted in the middle. Male: segment 1 of front tarsi very wide basally, with a longitudinal ridge ventrally (Fig. 212); the inner front claw distinctly shorter, more strongly curved and wider than the outer; segment 1 of mid tarsi usually somewhat incised ventrally; penis long, with a narrow, apically rounded distal part, and with an extension or "hood" and a denticle near apex dorsally; ventral edge of right paramere hardly dilated distally, with a large, dense tuft of hairs in the middle, and a small one at apex; genital sclerites, vide Fig. 204. Female: elytra without discernable micro-punctuation; tergal halves IX much longer than gonocoxae; genital sclerites, vide Fig. 254.

Distribution. All of Denmark. – Mainly in the south of Sweden, but reaching Nb. along the coast. – Norway: only in the south and southeast: Ø, AK, VE, VAy. – Finland: in all coastal districts, and in Ta and Tb. – Adjacent parts of the USSR: only Vib. – Mainly in central Europe, north to Britain, Fennoscandia and parts of European USSR, south to Spain, France (including Corsica), Austria and Hungary; eastwards to northeastern Siberia (Yakutian ASSR).

Biology. Chiefly in base-rich stagnant or slowly running water, and often found in larger ponds and lakes; also in brackish water, especially in the north. This is a characteristic species of euproductive stagnant waters (Seeger, 1971c). The larva, and to a rather large extent also the adult, feeds on filamentous algae (e.g. *Cladophora*) (Seeger, 1971b). The life-cycle has been worked out in detail by Falkenström (1926) and Seeger (1971a, b, c), who also provide descriptions of the larva. The life-cycle under natural conditions seems largely to be as described for *flavicollis.* This species is known to fly (Jackson, 1973).

Family Hygrobiidae

Morphology of the adult (Figs 255-258, 263-265).

Mainly diagnostic characters are given below, as the family does not occur in Fennoscandia and Denmark.

Body length 8.0-11.0 mm. Oval, dorsal and ventral sides very convex.

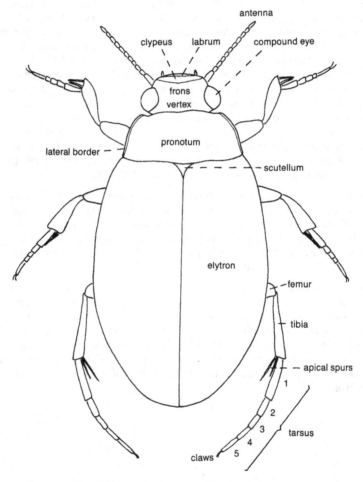

Fig. 255. *Hygrobia hermanni* (F.), dorsal view of ♀.

Head fairly large, prognathous. Compound eyes not divided into ventral and dorsal portions, strongly protruding. Labrum short, transverse. Clypeus transverse, separated from frons by a suture. Other dorsal sclerites of head capsule fused without separating sutures. Antennae filiform, almost glabrous, 11-segmented; basal segment rather long. Mouthparts largely as described for the Haliplidae; distal segment of galea with a bifid apex.

Anterior edge of prothorax with a fringe of long hairs. Pronotum bordered laterally; posterior edge produced in the middle, but not covering the scutellum. Prosternal apophysis well developed, rather narrow, projecting over the small mesosternum, and

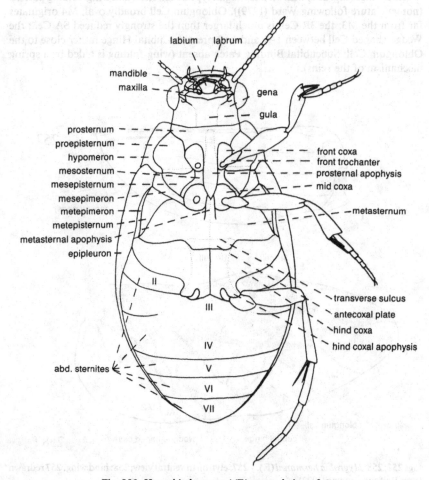

Fig. 256. *Hygrobia hermanni* (F.), ventral view of ♀.

137

reaching the metasternal apophysis. Proepisterna well developed. Mesepimeron and mesepisternum both reach the elytral epipleuron. Metasternum rather short, laterally not reaching the epipleura; divided by a transverse suture, separating an antecoxal plate from the remaining part of the sclerite. Metepisternum not reaching the mid coxal cavities, and separated from the epipleuron by the narrow metepimeron.

The elytra cover the abdomen when at rest. They are bordered laterally and densely punctured. Ventral side with a presutural row of transverse furrows distally (Fig. 257); this is the pars stridens of a stridulatory apparatus which has the posterior edge of abdominal sternite VII as the plectrum.

Hind wings (Fig. 258) with an adephagan type of venation and wing-folding pattern (nomenclature following Ward (1979)). Oblongum Cell broadly oval; M4 originates far from the M3; the 3R Cell is much larger than the strongly reduced SA Cell; the Wedge-shaped Cell between 1A1 and 1A2 present; Cubital Hinge rather close to the Oblongum Cell; Subcubital Binding Patch absent (wing-folding is aided by a spring mechanism of the veins).

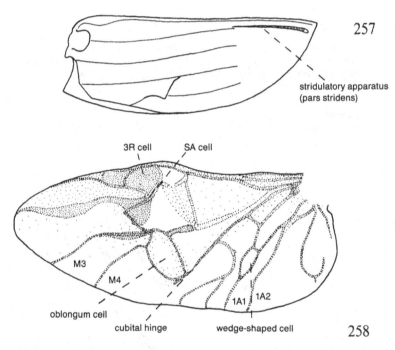

Figs 257, 258. *Hygrobia hermanni* (F.). – 257: elytron in ventral view; 258: hind wing. 257 redrawn from Balfour-Browne (1922).

138

Front and mid legs with a movable coxa, not particularly modified. Hind legs longer, modified for swimming. Hind coxae rather short, immovable, laterally reaching, or almost reaching, the epipleura; their median portion is somewhat elevated, but does not form the characteristic plates of the Noteridae; their apophyses cover only the basal parts of the hind trochanters. All tarsi 5-segmented, with the penultimate segment comparatively small in front and mid tarsi. Males have the basal three or four segments of front and mid tarsi dilated, provided with sucker-hairs ventrally.

Abdominal sternites II-VII visible from below. Gonocoxosternites not fused, invaginated. Tergites I-VIII and pleurites hidden under the resting elytra. Sternites II-IV connate, both together and with the meso- and metathoracic sclerites; the suture between sternites III and IV vanishes in the middle.

Genital sclerites (Figs 263-265) resemble those of the Haliplidae, but are more symmetrical bilaterally in males, with parameres of subequal shape and size.

Most of body surface densely punctured. Pronotum and elytra in the European species with a fine polygonal micro-reticulation. Body surface largely glabrous.

Morphology of the larva (Figs 259-261)

Elongate with rather wide head and thorax, body tapering behind. Body with ventral gill-filaments on the thoracic segments and on abdominal segments I-III (Fig. 261). Mandibles curved, slender and sharply pointed (Fig. 260); without a retinaculum, and without a suctorial channel or tube. Antennae 4-segmented with a very short distal segment. Legs slender, with two tarsal claws. Abdomen with 8 segments easily visible; segment VIII produced over rudimentary segment IX to a very long and narrow spine. Urogomphi very long.

Distribution

The family comprises only one genus, *Hygrobia*, with one West-Palearctic, one Chinese and three Australian species.

Biology

Mainly in stagnant water. The adults are fairly good swimmers (hind legs moved alternately), and are largely carnivorous. Below the surface they carry an air-supply below the elytra; this supply they renew from time to time by going to the surface, and there separating the abdomen from the elytra posteriorly. The beetle is capable of making quite loud "squeaks", when disturbed. The larvae are predaceous; they do not come to the surface to respire, as gas-exchange takes place largely through the gill-filaments.

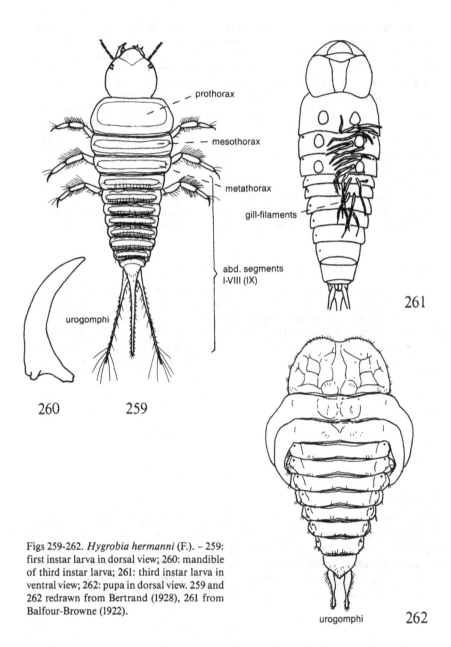

prothorax

mesothorax

metathorax

gill-filaments

abd. segments
I-VIII (IX)

261

urogomphi

260   259

Figs 259-262. *Hygrobia hermanni* (F.). – 259:
first instar larva in dorsal view; 260: mandible
of third instar larva; 261: third instar larva in
ventral view; 262: pupa in dorsal view. 259 and
262 redrawn from Bertrand (1928), 261 from
Balfour-Browne (1922).

urogomphi   262

Genus *Hygrobia* Latreille, 1804

Hydrachna Fabricius, 1801, Syst. Eleuth. 1: 255 (in part); preocc., *nec* Müller, 1776 (Acarina); see note.
Type species: not designated.
Hygrobia Latreille, 1804, Nouv. Dict. Hist. nat. 24: 139; emend. of *Hygriobia;* see note.
Type species: *Dytiscus hermanni* Fabricius, 1775 [= *Hygrobia hermanni* Fabricius, 1775)], by monotypy.
Paelobius Schönherr, 1808, Syn. Insect. 1 (2): 27.
Type species: *Dytiscus tardus* Herbst, 1779 [= *Hygrobia hermanni* (Fabricius, 1775)], by monotypy.
Pelobius Erichson, 1832, Gen. Dytisc.: 19, 45; unjustif. emend. of *Paelobius* Schönherr, 1808.
Type species: *Dytiscus hermanni* Fabricius, 1775 [= *Hygrobia hermanni* (Fabricius, 1775)], by monotypy.

The diagnosis of the genus is the same as for the family, given above.

Systematic notes: the remaining species originally included in genus *Hydrachna* Fabricius, 1801, all belong to genus *Hyphydrus* Illiger, 1802 (Biström, 1982).

The spelling *"Hygrobia"* is valid for the genus created by Latreille (1804). This was fixed by the International Commission on Zoological Nomenclature (1954: opinion 280).

Hygrobia hermanni (Fabricius, 1775)
Figs 255-265; pl. 1: 5.

Dytiscus hermanni Fabricius, 1775, Syst. Ent.: 232; emend. of *Herrmanni;* see note.
Dytiscus tardus Herbst, 1779, Schrift. Naturforsch. Ges. Berlin 4: 318.

8.0-11.0. Dorsal side yellowish or reddish with black markings around the eyes, along the anterior and posterior margins of pronotum, and forming a large, common, irregular spot (sometimes dissolved into stripes) on the elytra; rarely with pronotum and elytra almost completely black. Ventral side and legs largely yellowish or reddish, with parts of the head, thorax, and posterior abdominal sternites darkened, sometimes black laterally. Punctuation dense, fine on the head, and with both fine and coarse punctures on pronotum and the elytra. Most of the ventral surface with very dense, often confluent punctuation, giving it a granular appearance. Pronotum, elytra, and often hind coxae with a fine, isodiametric micro-reticulation, visible at low magnification, × 40. Each elytron with four rows of punctures which are sparsely provided with long hairs. Male: penis pointed apically, slightly asymmetrical; genital sclerites, vide Figs 263, 264. Female: genital sclerites, vide Fig. 265, rather stout.

Distribution. Not in Denmark and Fennoscandia. – In Europe mainly Mediterrane-

141

an, but reaching Britain, the Netherlands, German F. R. and D. R. in the north; Morocco, Tunisia, Algeria, USSR (only western Ukraine).

Biology. Mainly in stagnant water, in the north especially in silt and detrius ponds. The adult is largely carnivorous. The life-cycle from egg to adult lasts about 9 to 15 weeks, and was worked out in detail by Balfour-Browne (1922): eggs are deposited from late March to June on the surface of submerged vegetation. The larva (Fig. 259) passes through three instars, and mainly preys on *Tubifex*. The pupal stage is passed in the soil out of the water; it lasts about 16 days. The pupa (Fig. 262) is whitish, without setiferous tubercles, and with well developed urogomphi; it is furthermore characterized by the shape of the hind coxal apophyses, resembling those of the adult. The larva and the pupa were first described by Schiødte (1872). The adult hibernates in the water. In the Pyrenees the species has been recorded from an altitude of 2000 m. (Bertrand, 1972).

Note on synonymy. The original spelling, *"Herrmanni"*, has been rejected by the International Commission on Zoological Nomenclature (1954: opinion 280).

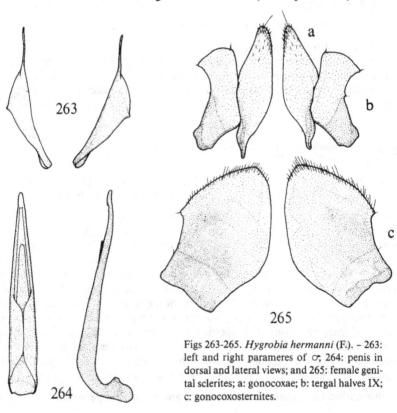

Figs 263-265. *Hygrobia hermanni* (F.). – 263: left and right parameres of ♂; 264: penis in dorsal and lateral views; and 265: female genital sclerites; a: gonocoxae; b: tergal halves IX; c: gonocoxosternites.

FAMILY NOTERIDAE

Morphology of the adult (Figs 266-269).

The morphology of both adults and larvae in many respects resemble that of the Dytiscidae (treated in a future volume), and mainly diagnostic characters are therefore given below.

Body length varies from less than 1 mm. to about 8 mm. (3.5-5.0 mm. in the Fennoscandian species). Dorsal side very convex, ventral side flattened.

Head small, prognathous. Compound eyes not divided into ventral and dorsal portions, hardly protruding. Labrum transverse, rather short. Other dorsal sclerites of head capsule fused without separating sutures. Antennae filiform, sometimes with widened segments, 11-segmented. Mouthparts largely as described for the Haliplidae; distal segment of labial palpus much longer than the preceding segments; mandibles with bifid apex.

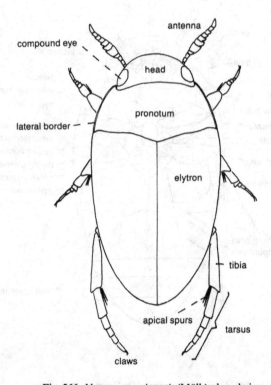

Fig. 266. *Noterus crassicornis* (Müll.), dorsal view.

Pronotum narrowed anteriorly, bordered laterally; posterior margin produced in the middle, covering the scutellum. Prosternal apophysis well developed, widened posteriorly, projecting over the small mesosternum, and reaching the metasternal apophysis. Proepisterna well developed. Mesepimeron separates mesepisternum from the elytral epipleuron. Metasternum not very large, laterally not reaching the elytral epipleura, not divided by a transverse suture. Metepisternum does not reach the mid coxal cavities. Metepimera not distinct from below.

Elytra often rather strongly attenuated behind, covering the abdomen. Elytra bordered laterally; dorsally often with punctures which are sometimes arranged in rows.

Hind wings (Fig. 268) with an Adephagan type of venation and wing-folding pattern (nomenclature following Ward (1979)). Oblongum Cell often almost square; M4 originates far from the M3; the Wedge-shaped Cell between 1A1 and 1A2 present; Cubital Hinge very close to the Oblongum Cell; Subcubital Binding Patch present.

Front and mid legs with a movable coxa, rather short, modified for crawling or bur-

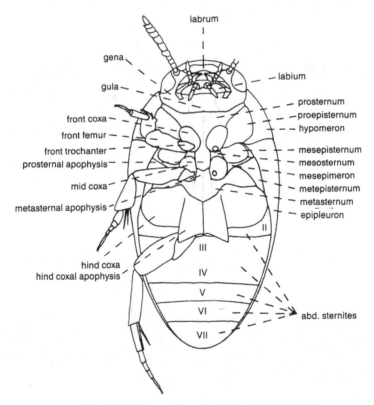

Fig. 267. *Noterus clavicornis* (De Geer), ventral view.

rowing. Hind legs longer, modified for swimming. Hind coxae large, immovable, laterally reaching the epipleura; their median portion is somewhat elevated, forming a characteristic plate which usually covers the bases of the hind trochanters and the hind femora, plus small portions of the hind coxae and abdominal sternites II-III. All tarsi distinctly 5-segmented.

Abdominal sternites II-VII visible from below; gonocoxosternites not fused, invaginated. Tergites I-VIII and pleurites hidden by the elytra. Sternites II-IV connate, both together and with meso- and metathoracic sclerites; the suture between sternites III and IV often indistinct.

Genitalia much as described for the Haliplidae.

Dorsal surface of body sometimes finely micro-reticulated (Fig. 269), and often punctured; largely glabrous.

Morphology of the larva (Figs 270-273)

Elongate, body without micro-tracheal gills or gill filaments. Mandibles stout, suctorial tube or channel absent, retinaculum present (Fig. 271). Antennae short, 4-segmented. Legs short, with two tarsal claws (Figs 272, 273). Abdomen with 8 visible segments; segment VIII conical, more or less produced over the rudimentary segment IX to a narrow apex bearing a pair of spiracles. Urogomphi present, often short.

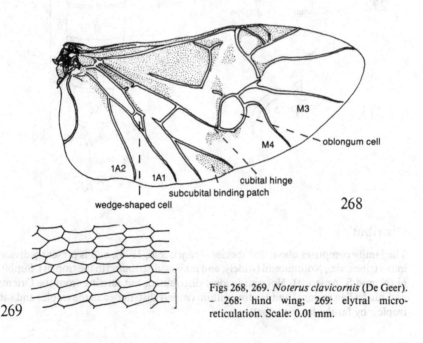

M3

M4

oblongum cell

1A2

1A1

cubital hinge

subcubital binding patch

wedge-shaped cell

268

269

Figs 268, 269. *Noterus clavicornis* (De Geer). — 268: hind wing; 269: elytral micro-reticulation. Scale: 0.01 mm.

145

Figs 270, 271. *Noterus clavicornis* (De Geer). – 270: third
instar larva in dorsal view; 271: larval mandible.
Figs 272, 273. Hind leg of third instar larvae of 272: *Noterus clavicornis* (De Geer) and 273: *N. crassicornis* (Müll.).

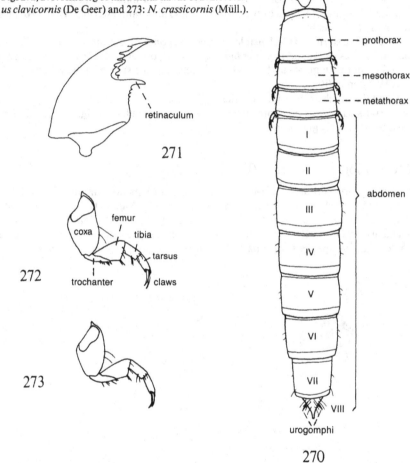

retinaculum

271

femur
coxa
tibia
tarsus
272
trochanter
claws

273

prothorax
mesothorax
metathorax
I
II
abdomen
III
IV
V
VI
VII
VIII
urogomphi

270

Distribution

The family comprises about 260 species (Franciscolo, 1979), and is presently divided
into 4 tribes, viz., Notomicrini (widely, and mainly, distributed in the tropics), Suphisini (N. and S. America), Noterini (widely distributed, but missing from the Oriental
and Australian regions) and Hydrocanthini (widely distributed in the tropics and subtropics; by far the largest tribe).

146

Biology

The species of Noteridae are mainly found in stagnant water, often among dense vegetation, and particularly in places with much decaying material. The adults are fairly good swimmers, but mainly crawl among the vegetation (hind legs moved simultaneously while swimming). They are probably largely carnivorous. Below the surface they carry an air-supply below the elytra; this supply they renew from time to time by going to the surface, and there separating the abdomen from the elytra posteriorly.

Not much is known about the immature stages in general. The life-cycle of the Fennoscandian species of *Noterus* has been worked out, and is described below. These larvae are known for their burrowing habits, but the known larvae of other genera live among submerged vegetation like many dytiscid larvae, which they resemble morphologically (Bertrand, 1972), and almost nothing is known about their biology.

TRIBE NOTERINI

Anterior margin of pronotum bordered by a row of punctures, sometimes connected by a furrow. Elytra without dense, evenly dispersed punctuation; punctures more scattered, often more or less arranged in rows. The punctures are often rather coarse, sometimes granular or squamiform. Apex of prosternal apophysis rounded or truncate. Hind coxal apophyses separated behind by a triangular space. Tibiae with two apical spurs. Front tibia with an extension covering the base of the tarsus, and with one of the apical spurs very long and curved (Fig. 274). Posterior apical angle of hind tibia without a dense fringe of hairs.

The tribe contains about 25 species (Franciscolo, 1979) in 6 genera, viz., *Siolius* J. Balfour-Browne (S. America), *Renotus* Guignot (Africa), *Pronoterus* Sharp (N. and S. America), *Mesonoterus* Sharp (Guatemala), *Synchortus* Sharp (Africa), and *Noterus* Clairville (Palearctic region).

Genus *Noterus* Clairville, 1806

Noterus Clairville, 1806, Ent. Helv. 2: 222.
Type species: *Dytiscus crassicornis* Müller, 1776 (attributed to Fabricius) [= *Noterus crassicornis* (Müller, 1776)], by monotypy.

Medium-sized species. Body elongate, widest in front of the middle, dorsally strongly convex, ventrally flattened, largely reddish or brownish. Head with polygonal microreticulation, pronotum and elytra with longitudinal rows of very fine, transverse meshes, visible at high magnification, ×150 (Fig. 269). Antennae widened, especially in males. Pronotum with a wide lateral border. Elytral punctuation rather fine and sparse anteriorly, in the Fennoscandian species stronger and denser posteriorly. Prosternal apophysis widened distally with apex forming an obtuse angle or rounded. Male: parts of the ventral side often darkened; anterior legs widened, femur, ventrally

147

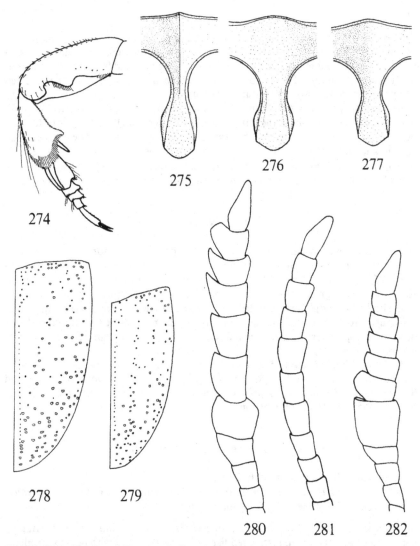

Fig. 274. Male front leg of *Noterus clavicornis* (De Geer).
Figs 275-277. Median portion of prosternum of *Noterus* spp. – 275: *N. clavicornis* (De Geer); 276: *N. crassicornis* (Müll.) ♂; and 277: same species ♀.
Figs 278, 279. Right elytron of 278: *Noterus clavicornis* (De Geer) and 279: *N. crassicornis* (Müll.).
Figs 280-282. Antennae of *Noterus* spp. – 280: *N. clavicornis* (De Geer) ♂; 281: same species ♀; 282: *N. crassicornis* (Müll.) ♂.

148

near the middle, and tibia, distally along the inner edge, with blunt dilations (Fig. 274); segments 1-3 of front tarsi and 1-2 of mid tarsi dilated, provided with suckers ventrally.

The genus is widespread in the Palearctic region. 7 species have been described, two of which are known from northern Europe.

The species are chiefly found in permanent stagnant bodies of water with much decaying vegetation. The immature stages all live below the surface of the water, near the roots of various semi-aquatic monocotyledons.

Key to species of *Noterus*

1   Anterior half of prosternum in both sexes medially convex with a longitudinal ridge (Fig. 275). Elytral punctures usually not arranged in very distinct rows (Fig. 278). Male antennal segment 5 about as long as segment 6 (Fig. 280). Length: 4.0-5.0 mm .... 35. *clavicornis* (De Geer)
–   Anterior half of pronotum in females medially convex (Fig. 277), in males flattened (Fig. 276), in both without a longitudinal ridge. Elytral punctures usually forming longitudinal rows anteriorly (Fig. 279). Male antennal segment 5 about twice as long as segment 6 (Fig. 282). Length: 3.5-4.2 mm ....
.... 36. *crassicornis* (Müller)

35. *Noterus clavicornis* (De Geer, 1774)
Figs 267-272, 274, 275, 278, 280, 281, 283, 285, 287, 288.

Dytiscus clavicornis De Geer, 1774, Mém. Hist. Ins. 4: 402.
Dytiscus capricornis Herbst, 1784, *in* Füessly: Arch. Insektengesch. 5: 125 (in part).
Dytiscus semipunctata Fabricius, 1792, Ent. Syst. 1: 199.
Dytiscus sparsus Marsham, 1802, Ent. Brit. 1: 430.
Noterus crassicornis, sensu auctt.; misident., *nec* (Müller, 1776).

4.0-5.0 mm. Dorsal side brownish or reddish (less reddish than in *crassicornis*), with the posterior margin of head, a median spot on pronotum, and the elytra somewhat darker. Males with ventral side and basal parts of legs dark, almost black; females with the ventral side and legs brownish or reddish. Elytral punctures not arranged in very distinct rows (Fig. 278). Anterior half of prosternum in both sexes medially convex with a longitudinal ridge (Fig. 275). Elevated portion of hind coxa with fine and rather dense punctuation. Male: antennal segments 5-8 both very long and wide (Fig. 280); penis without an angular dilation of the ventral edge near the base; genital sclerites, vide Fig. 283. Female: antennal segments hardly widened (Fig. 281); tergal half IX much longer than gonocoxa; genital sclerites, vide Fig. 285.

Distribution. All of Denmark. – Mainly in the south of Sweden, north to Boh., Ög., Vg. and Sdm. – Norway: only in the southeast: Ø, AK, VE. – In the southwest of Fin-

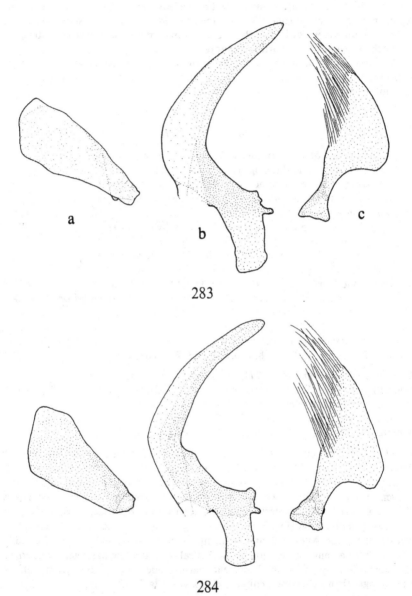

a

b

c

283

284

Figs 283, 284. Male genital sclerites of *Noterus* spp.; a: left paramere; b: penis; c: right paramere. – 283: *N. clavicornis* (De Geer) and 284: *N. crassicornis* (Müll.).

land: Al, Ab, N, St. – Not in adjacent parts of the USSR. – Europe north to Britain and Fennoscandia; central and southern parts of European USSR; Syria, Lebanon, Turkey, Iran, Iraq, the Caucasus, Transcaspia, Turkestan, eastwards to Kashmir, Mongolia and China.

Biology. In base-rich, often polyproductive bodies of stagnant water. Mainly in non-temporary waters. Many Fennoscandian records are from brackish water or from localities very near the sea. The adult is usually found among submerged decaying vegetation, and is probably largely carnivorous. The egg is elongate and whitish. One egg has been discovered in late May, deposited under a decaying leaf-base on *Scirpus*, but whether this is the usual place for oviposition is not certain.

The larva (Fig. 270) is elongate, subcylindrical, and of a pale yellowish colour; first instar larvae have a darker transverse band across most segments which is more or less missing in the two following instars. Young larvae have been obtained in late June, and full grown ones in the beginning of August. The larva burrows among the roots of various semi-aquatic monocotyledons such as *Typha, Scirpus, Iris, Juncus, Alisma, Spar*-

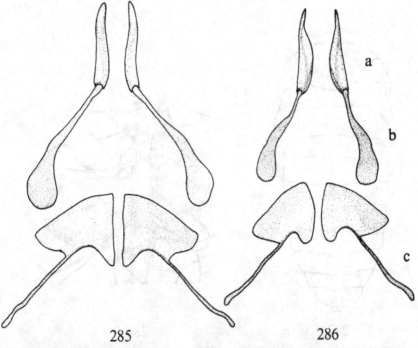

285   286

Figs 285, 286. Female genital sclerites of *Noterus* spp.; a: gonocoxae; b: tergal halves IX; c: gonocoxosternites. – 285: *N. clavicornis* (De Geer) and 286: *N. crassicornis* (Müll.).

151

ganium and some grasses. Newly caught larvae usually have the intestine filled with a debris-like material, but in aquaria they feed vigorously on dead insects and pieces of meat which are held by the legs, bitten to pieces by the mandibles, and ingested through the mouth. They do not come to the surface to respire unless forced, as their gas-exchange takes place largely by the penetration of roots with the pointed tip of abdominal segment VIII. The larvae are very agile when burrowing, but otherwise rather poor crawlers, partly moving by characteristic sideward movements of the body.

The larva was first described by Meinert (1901), and further details on its biology and morphology have been provided by F. Balfour-Browne & J. Balfour-Browne (1940) and other authors. Third instar larvae of this species may be separated from those of *crassicornis* by the comparatively wider femora (Fig. 272). Prior to pupation the larva produces a pupal cocoon (Fig. 288) by binding particles of the surrounding substratum with a glandular secretion. This cocoon is filled with air through a penetrated root to which it is attached. The pupa (Fig. 287) is of a whitish colour, lacks setiferous tubercles, and has the urogomphi strongly reduced. It is furthermore characterized by the elevated hind coxal plates and the shape of the male antennae, also seen in the adult. Pupal cocoons and young adults have been obtained in August. The pupa and the cocoon were first figured by F. Balfour-Browne & J. Balfour-Browne

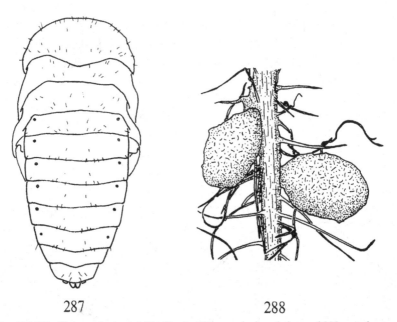

287   288

Figs 287, 288. *Noterus clavicornis* (De Geer) – 287: pupa in dorsal view and 288: pupal cocoons on a root. – 288 redrawn from F. & J. Balfour-Browne (1940).

152

(1940). The adult hibernates below the surface of the water. Most British specimens examined by Jackson (1973) were not capable of flying due to skeletal and muscular reduction. The species has been recorded from an altitude of 1450 m. in Mongolia (Guéorguiev, 1965).

36. *Noterus crassicornis* (Müller, 1776)
    Figs 266, 273, 276, 277, 279, 282, 284, 286; pl. 1: 6.

Dytiscus crassicornis Müller, 1776, Zool. Dan. Prodr.: 72.
Dytiscus capricornis Herbst, 1784, *in* Füessly: Arch. Insektengesch. 5: 125 (in part).
Noterus Geeri Leach, 1817, Zool. Miscellany 3: 71.
Noterus clavicornis, sensu auctt.; misident., *nec* (De Geer, 1774).

3.5-4.2 mm. Resembling the preceding species, but differing as follows: colour more reddish, median spot on pronotum absent or almost so. Ventral side in both sexes largely reddish, in males with lateral parts of head and prothorax and distal parts of front and mid femora dark, almost black. Elytral punctures usually forming longitudinal rows anteriorly (Fig. 279). Anterior half of prosternum in both sexes without a median longitudinal ridge, flattened medially in males (Figs 276, 277). Elevated portion of hind coxa with fine and rather sparse punctuation. Male: antennal segment 5 both very long and wide (Fig. 282); penis with an angular dilation of the ventral edge near the base; genital sclerites, vide Fig. 284. Female: tergal half IX only slightly longer than gonocoxa; genital sclerites, vide Fig. 286.

Distribution. Common in all of Denmark. - Mainly in the southern parts of Sweden, but records as far north as Nb. - Norway: known from a number of southeastern districts; also Ry in the southwest. - Widely distributed in the south of Finland, recorded from all districts northwards to ObS and Ok. - Adjacent parts of the USSR: Vib, Kr. - In Europe north to Britain, Fennoscandia and European parts of the USSR, south to the Pyrenees, Italy, and the Balkans; Turkey, Iran, eastwards to Siberia and China.

Biology. Chiefly found in permanent bodies of stagnant water. It may be found in a large variety of habitats, but usually not in brackish water (Jeppesen, 1980). Otherwise adults, larvae and pupae occur as described for *clavicornis*. The third instar larvae may be separated from those of *clavicornis* by the comparatively narrower femora (Fig. 273). The larva was briefly described by Bertrand (1950), and further details on the immature stages are presented by Ruhnau (1985). All European adults examined by Jackson (1973) had more or less reduced hind wings, and were thus probably not capable of flight. The larva is sometimes infected with an endoparasitic species of Gordiidae which leaves the larva shortly before its pupation.

153

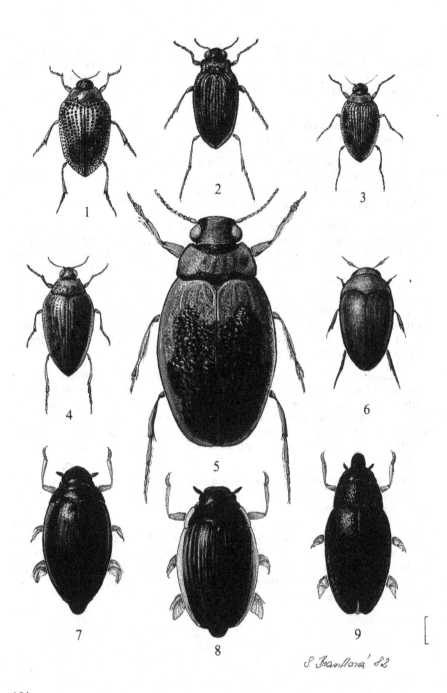

1

2

3

4

5

6

7

8

9

S. Brantlova '82

154

Catalogue

		Germany	G. Britain	SJ	EJ	WJ	NWJ	NEJ	F	LFM	SZ	NWZ	NEZ	B	Sk.	Bl.
Gyrinus minutus F.	1	●	●	●	●	●	●	●	●	●	●			●	●	●
G. opacus Sahlbg.	2	●														
G. aeratus Stph.	3	●	●	●	●	●	●	●	●				●	●	●	●
G. pullatus Zaits.	4															
G. marinus Gyll.	5	●	●	●	●	●	●	●	●	●	●	●	●	●	●	●
G. colymbus Er.	6	●							●							
G. distinctus Aubé	7	●	●	●	●	●	●	●		●	●	●			●	●
G. suffriani Scriba	8	●	●	●	●	●				●			●	●		●
G. natator (L.)	9	●	●						●	●			●	●		●
G. substriatus Stph.	10	●	●	●	●	●	●	●	●	●	●	●	●	●	●	●
G. caspius Mén.	11	●	●	●	●	●	●	●	●	●	●	●	●	●	●	●
G. paykulli Ochs	12	●	●	●	●	●	●	●	●	●	●	●	●	●	●	●
G. urinator Ill.		●	●													
Aulonogyrus concinnus (Klug)		●														
A. striatus (F.)			?													
Orectochilus villosus (Müll.)	13	●	●	●	●	●	●	●	●			●	●	●	●	●
Brychius elevatus (Panz.)	14	●	●	●	●	●	●	●	●					●		
Peltodytes caesus (Dft.)	15	●	●	●	●				●	●	●	●	●	●	●	●
Haliplus lineatocollis (Marsh.)	16	●	●	●	●	●	●	●	●	●	●	●	●	●	●	●
H. laminatus (Schall.)	17	●	●					●			●	●	●	●	●	●
H. flavicollis Sturm	18	●	●	●	●		●	●			●	●	●	●	●	●
H. mucronatus Stph.		●	●													
H. fulvus (F.)	19	●	●				●		●		●	●	●		●	●
H. variegatus Sturm	20	●	●	●	●	●	●	●	●	●	●	●	●	●	●	●
H. confinis Stph.	21	●	●	●	●	●	●	●	●	●	●	●	●	●	●	●
H. varius Nicol.	22	●	●													
H. obliquus (F.)	23	●	●	●		●			●	●	●	●	●		●	●
H. apicalis Thoms.	24	●	●	●		●			●	●	●	●	●	●	●	●
H. furcatus Seidl.	25	●	●			●				●	●	●		●	●	●
H. fulvicollis Er.	26	●		●	●							●		●		●
H. fluviatilis Aubé	27	●	●	●	●			●	●	●	●	●		●	●	●
H. lineolatus Mann.	28	●	●	●	●	●	●	●	●	●	●	●		●	●	●
H. interjectus Lindbg.	29															
H. sibiricus Motsch.	30															
H. wehnckei Gerh.	31	●	●	●	●	●	●	●	●	●	●	●	●	●	●	●
H. ruficollis (De Geer)	32	●	●	●	●	●	●	●	●	●	●	●	●	●	●	●
H. heydeni Wehncke	33	●	●	●	●	●	●	●	●	●	●	●	●	●	●	●
H. immaculatus Gerh.	34	●	●	●	●	●	●	●	●	●	●	●	●	●	●	●

SWEDEN

	Hall.	Sm.	Öl.	Gtl.	G. Sand.	Ög.	Vg.	Boh.	Dlsl.	Nrk.	Sdm.	Upl.	Vstm.	Vrm.	Dlr.	Gstr.	Hls.	Med.	Hrj.	Jmt.	Äng.	Vb.	Nb.	Ås. Lpm.	Ly. Lpm.	P. Lpm.	Lu. Lpm.	T. Lpm.
1	●	●	●	●	●	●	●	●	●	●	●	●	●	●	●	●	●	●	●	●	●	●	●	●	●	●	●	●
2						●	●				●	●	●		●	●		●		●	●	●	●	●	●	●	●	●
3	●	●	●			●	●	●	●	●	●	●	●	●	●	●	●	●		●	●	●	●	●	●	●	●	●
4																							●					
5	●	●	●	●			●	●	●			●	●	●	●	●				●	●	●	●	●	●	●	●	●
6																												
7	●	●			●		●	●	●			●	●	●	●		●											
8	●	●	●	●	●		●	●	●			●	●	●	●		●											
9		●	●	●	●	●	●	●	●	●	●	●	●	●		●		●		●	●		●	●				
10	●	●	●	●	●	●	●	●	●	●	●	●	●	●	●		●	●			●	●	●					
11	●	●									●																	
12	●	●	●	●			●	●				●		●			●		●			●						
13	●	●	●			●	●	●	●	●	●	●	●	●	●		●											
14	●	●				●	●				●	●		●						●	●	●	●				●	●
15		●	●																									
16	●		●	●			●																					
17																												
18	●	●	●	●			●	●	●	●	●	●			●	●		●	●		●	●	●					
19	●	●	●	●		●	●	●	●	●	●	●	●				●	●	●	●	●	●	●					
20	●					●	●					●		●									●					
21	●	●	●	●			●	●				●	●		●		●	●	●	●	●	●	●	●		●		●
22																												
23		●	●	●		●						●	●								●							
24	●	●	●	●				●																				
25			●	●			●																					
26	●	●	●	●			●	●				●	●	●														
27			●			●																						
28	●	●	●	●			●	●	●	●	●	●				●	●		●	●	●	●		●	●	●	●	
29																												
30																							●					●
31	●			●			●				●				●	●				●	●	●	●					
32	●	●	●	●	●		●	●	●	●	●	●	●	●	●	●	●	●	●	●	●	●	●					
33	●	●	●	●			●	●	●	●	●	●	●	●	●	●	●				●							
34	●	●	●	●	●		●	●	●	●	●	●	●					●	●		●	●	●					

157

		Ø+AK	HE (s+n)	O (s+n)	B (ø+v)	VE	TE (y+i)	AA (y+i)	VA (y+i)	R (y+i)	HO (y+i)	SF (y+i)	MR (y+i)	ST (y+i)	NT (y+i)	Ns (y+i)
Gyrinus minutus F.	1	●	●	●	●	●	●	◐		●	●	◐		◐	●		◐●
G. opacus Sahlbg.	2	●	●	●	◐		◐			●	●	◐	◐	◐		◐●
G. aeratus Stph.	3	●	●	◐	●	●	◐	◐		◐	◐	●	◐	◐		◐	◐
G. pullatus Zaits.	4															
G. marinus Gyll.	5	●			◐		●	◐	◐	◐	◐					
G. colymbus Er.	6															
G. distinctus Aubé	7								◐							
G. suffriani Scriba	8					◐										
G. natator (L.)	9	●														
G. substriatus Stph.	10	●	●	●	●	●	●	◐		◐	◐	◐	●	●	◐	◐	●	◐		◐
G. caspius Mén.	11						◐									
G. paykulli Ochs	12	◐							◐	◐	◐					
G. urinator Ill.																
Aulonogyrus concinnus (Klug)																
A. striatus (F.)																
Orectochilus villosus (Müll.)	13			◐	●			◐	◐							
Brychius elevatus (Panz.)	14	◐◐														
Peltodytes caesus (Dft.)	15															
Haliplus lineatocollis (Marsh.)	16															
H. laminatus (Schall.)	17															
H. flavicollis Sturm	18	◐◐			◐	●	◐									
H. mucronatus Stph.																
H. fulvus (F.)	19	●	●	●	●	●	●	●	◐	◐	◐	●	◐	●	●	●	●
H. variegatus Sturm	20	◐				●										
H. confinis Stph.	21	●		◐					◐				●		●
H. varius Nicol.	22															
H. obliquus (F.)	23															◐
H. apicalis Thoms.	24	●		◐			◐									
H. furcatus Seidl.	25															
H. fulvicollis Er.	26	◐														
H. fluviatilis Aubé	27															
H. lineolatus Mann.	28	◐◐	●	◐					◐				◐		
H. interjectus Lindbg.	29															
H. sibiricus Motsch.	30															
H. wehnckei Gerh.	31					◐		●	◐	◐	◐●	●	●	◐	◐
H. ruficollis (De Geer)	32	●	●	◐	●	●	●	◐	◐	◐	●	◐	◐	●	●	◐	●
H. heydeni Wehncke	33	●		◐	◐		◐	◐	◐						
H. immaculatus Gerh.	34	●				●			◐						

158

	Nn (ø + v)	TR (y + i)	F (v + i)	F (n + ø)	Al	Ab	N	Ka	St	Ta	Sa	Öa	Tb	Sb	Kb	Om	Ok	Ob S	Ob N	Ks	LkW	LkE	Lc	Li	Vib	Kr	Lr
1																											
2																											
3																											
4																											
5																											
6																											
7																											
8																											
9																											
10																											
11																											
12																											
13																											
14																											
15																											
16																											
17																											
18																											
19																											
20																											
21																											
22																											
23																											
24																											
25																											
26																											
27																											
28																											
29																											
30																											
31																											
32																											
33																											
34																											

		Germany	G. Britain	SJ	EJ	WJ	NWJ	NEJ	F	LFM	SZ	NWZ	NEZ	B	Sk.	Bl.
Hygrobia hermanni (F.)		●	●													
Noterus clavicornis (De Geer)	35	●	●	●	●	●	●	●	●	●	●	●	●	●	●	●
N. crassicornis (Müll.)	36	●	●	●	●	●	●	●	●	●	●	●	●	●	●	●

		Ø+AK	HE (s+n)	O (s+n)	B (ø+v)	VE	TE (y+i)	AA (y+i)	VA (y+i)	R (y+i)	HO (y+i)	SF (y+i)	MR (y+i)	ST (y+i)	NT (y+i)	Ns (y+i)
Hygrobia hermanni (F.)																
Noterus clavicornis (De Geer)	35	●				●										
N. crassicornis (Müll.)	36	●	◐	◐	●	●	◐	◐		◐						

	Hall.	Sm.	Öl.	Gtl.	G. Sand.	Ög.	Vg.	Boh.	Dlsl.	Nrk.	Sdm.	Upl.	Vstm.	Vrm.	Dlr.	Gstr.	Hls.	Med.	Hrj.	Jmt.	Ång.	Vb.	Nb.	Ås. Lpm.	Ly. Lpm.	P. Lpm.	Lu. Lpm.	T. Lpm.
35	●	●	●	●		●	●	●			●																	
36	●	●	●	●		●	●	●	●	●	●	●	●	●	●	●	●	●					●					

	Nn (ø+v)	TR (y+i)	F (v+i)	F (n+ø)	Al	Ab	N	Ka	St	Ta	Sa	Oa	Tb	Sb	Kb	Om	Ok	ObS	ObN	Ks	LkW	LkE	Le	Li	Vib	Kr	Lr
35					●	●	●	●	●																		
36					●	●	●	●	●	●	●	●	●	●	●	●	●	●							●	●	

Literature

References to original descriptions are given under the respective taxa.

Andersen, J., 1970: The larvae of *Pelophila borealis* Paykull, *Nebria gyllenhali* Schnh. and *N. nivalis* Paykull (Coleoptera, Carabidae). – Astarte 3: 87-95.

Aubé, C., 1838: Hydrocanthares et Gyriniens *in* Dejean: Spécies Général des Coleóptères de la Collection de m. le Comte Dejean 6. – Paris.

Balfour-Browne, F., 1915: On the British species of *Haliplus*, Latreille, related to *Haliplus ruficollis*, De Geer, with some remarks upon *H. fulvicollis*, Erichson, and *H. furcatus*, Seidlitz. – Ann. Mag. nat. Hist. 15: 97-124.

– 1922: The life history of a water beetle: *Pelobius tardus* Herbst. – Proc. zool. Soc. Lond. (1922): 338-352.

– 1936a: Systematic notes upon British aquatic Coleoptera 11. – Entomologist's mon. Mag. 72: 68-77.

– 1936b: Systematic notes upon British aquatic Coleoptera 12. – Entomologist's mon. Mag. 72: 97-108.

– 1938: Systematic notes upon British aquatic Coleoptera 1. – London.

– 1940a: Concerning the three species *Haliplus lineolatus* Mannerh., *H. wehnckei* Gerh. and *H. heydeni* Wehncke (Col., Hydradephaga), with some remarks upon the orientation of the male sexual armature in the Hydradephaga. – Entomologist's mon. Mag. 76: 121-128.

– 1940b: British Water Beetles 1. – London.

– 1950: British Water Beetles 2. – London.

Balfour-Browne, F., & J. Balfour-Browne, 1940: An outline of the habits of the water-beetle, *Noterus capricornis* Herbst (Coleopt.). – Proc. R. ent. Soc. Lond. (A) 15: 105-112.

Balfour-Browne, J., 1945: The genera of the Gyrinoidea and their genotypes. – Ann. Mag. nat. Hist., Ser. 11, 12: 103-111.

Beier, M., 1929: Zur Kenntniss der Lebenweise von *Haliplus wehnckei* Gerh. – Z. Morph. Ökol. Tiere 14: 191-223.

Berg, K., 1948: Biological studies on the River Susaa. – Folia limnol. scand. 4: 1-318.

Bertrand, H., 1923: Larve et nymphe d'*Haliplus lineaticollis* Marsh. Anomalies chez les larves et nymphes des Coléoptères. – Annls Soc. ent. Fr. 93: 322-344.

– 1928: Les larves et nymphes des Dytiscides, Hygrobiides, Haliplides. – Encycl. ent. 10. Paris.

– 1942: Captures et élevages de larves de Coléoptères aquatiques 9. – Annls Soc. ent. Fr. 111: 67-74.

– 1950: Captures et élevages de larves de Coléoptères aquatiques 11. – Bull. Soc. ent. Fr. 55: 42-48.

– 1951: Captures et élevages de larves de Coléoptères aquatiques 13. – Bull. Soc. ent. Fr. 56: 75-80.

– 1972: Larves et nymphes des Coléoptères aquatiques du globe. – Paris.

Bertrand, H., & F. Vaillant, 1950: Observations biologiques sur les Gyrinides (Col.); la nymphe des *Aulonogyrus* Rég. – Bull. Hist. nat. Afrique du Nord 41: 15-19.

Biström, O., 1982: A revision of the genus *Hyphydrus* Illiger (Coleoptera, Dytiscidae). – Acta zool. fenn. 165: 1-121.

Biström, O., & H. Silfverberg, 1983: Additions and corrections to Enumeratio Coleopterorum Fennoscandiae et Daniae. – Notul. ent. 63: 1-9.

Bott, R. H., 1928: Beiträge zur Kenntniss von *Gyrinus natator substriatus* Steph. – Z. Morph. Ökol. Tiere 10: 207-306.

Brinck, P., 1944: Zweiter Beitrag zur Kenntnis kritischer Wasserkäfer-Arten. - Opusc. ent. 9: 149-154.

- 1955: A revision of the Gyrinidae (Coleoptera) of the Ethiopian region 1. - Lunds univ. Årsskr. n. f. 51 (16): 1-141.

Burmeister, E.-G., 1976: Der Ovipositor der Hydradephaga (Coleoptera) und seine phylogenetische Bedeutung unter besonderer Berücksichtigung der Dytiscidae. - Zoomorphologie 85: 165-257.

Böcher, J. (in press): The Coleoptera of Greenland. - Greenland Bio-science.

Crowson, R. A., 1955: The natural classification of the families of Coleoptera. - London.

Dolmen, D., & J. I. Koksvik, 1976: Haliplus obliquus Fabricius (Col., Haliplidae) ny for Norge. - Norw. J. Ent. 23: 203.

Erichson, W. F., 1837: Die Käfer der Mark Brandenburg 1. - Berlin.

Fabricius, J. C., 1787: Mantissa Insectorum, etc. 1. - Hafniae.

- 1801: Systema Eleutheratorum, etc. 1. - Hafniae.

Fairmaire, L., & Laboulbène, 1854: Faune entomologique Francaise, etc., Coléoptères 1. - Paris.

Falkenström, G., 1939: Revision des Haliplus lineolatus Mnh. und seiner systematischen Verwandten nebst einiger Neuheiten. Eine kritische Darstellung von morphologischen und biologischen Gesichtspunkten aus. - Ark. Zool. 32 A (6): 1-46.

Fall, H. C., 1921: The North American species of Gyrinus (Coleoptera). - Trans. Am. ent. Soc. 47: 269-306.

Fowler, W. W., 1887: The Coleoptera of the British Islands 1. - London.

Franciscolo, M. E., 1979: Coleoptera; Haliplidae, Hygrobiidae, Gyrinidae, Dytiscidae. - Fauna d'Italia 14. Bologna.

Girling, M. A., 1984: A Little Ice Age extinction of a water beetle from Britain. - Boreas 13: 1-4.

Geoffroy, E. L., 1762: Histoire abregée des insectes qui se trouvent aux environs de Paris 1. - Paris.

Gernet, C. von, 1868: Beiträge zur Käferlarvenkunde. - Horae Soc. ent. Ross. 5: 7-22.

Guéorguiev, V. B., 1965: 48. Haliplidae und Dytiscidae. Ergebnisse der zoologischen Forschungen von Dr. Z. Kaszab in der Mongolei (Coleoptera). - Reichenbachia 7: 127-134.

Guignot, F., 1928: Notes sur les Haliplus du groupe fulvus F. (Coleopt. Haliplidae). - Annls Soc. ent. Fr. 97: 133-151.

- 1930: Notes sur quelques Haliplus (Col. Haliplidae). Bull. Soc. ent. Fr. 35: 71-75.

- 1931-33: Les Hydrocanthares de France. - Toulouse.

- 1939: À propos du type du genre Haliplus (Col. Haliplidae). - Bull. Soc. ent. Fr. 43: 176.

- 1946: Haliplides, Dytiscides et Gyrinides du Haut-Atlas marocain. - Revue fr. Ent. 13: 185-187.

- 1947: Coléoptères Hydrocanthares. - Faune de France 48, Paris.

- 1955: Sur la systematique des Haliplus (Col. Haliplidae). - Mém. Soc. r. ent. Belg. 27: 289-296.

Hagenlund, G., 1984: Nye funn av akvatiske Coleoptera i Sør-Norge. - Fauna norv. (Ser. B.) 31: 104-105.

Hickman, J. R., 1931: Contribution to the biology of the Haliplidae. - Ann. ent. Soc. Am. 24: 129-142.

Holmen, M., 1979: Hvirvleren Gyrinus colymbus Er. fundet i Danmark (Coleoptera: Gyrinidae). - Ent. Meddr 47: 85.

- 1981: Status over Danmarks Haliplidae (Coleoptera) med bemærkninger om zoogeografi og autøkologi. - Ent. Meddr 49: 1-14.

Hope, F. W., 1838: The Coleopterist's manual 2: - London.

Huldén, L., 1983a: Distribution of Gyrinidae (Coleoptera) in Eastern Fennoscandia. - Notul. ent. 63: 81-85.

- 1983b: Laboulbeniales (Ascomycetes) of Finland and adjacent parts of the U.S.S.R. - Karste-

163

nia 23: 31-136.

Illiger, J. K. W., 1798: Verzeichnis der Käfer Preussens, entworf. von Kugelann, ausgearb. von Illiger. – Berlin.

Jackson, D. J., 1952a: Observations on the capacity for flight of water beetles. – Proc. R. ent. Soc. Lond. 27: 57-70.

– 1952b: *Haliplus obliquus* F. (Col., Haliplidae) in Fife, with notes on other water beetles and bugs from the same habitat. – Entomologist's mon. Mag. 88: 108.

– 1956: The capacity for flight of certain water beetles and its bearing on their origin in the western Scottish Isles. – Proc. linn. Soc. Lond. 167: 76-96.

– 1973: The influence of flight capacity on the distribution of aquatic Coleoptera in Fife and Kinross-shire. – Entomologist's Gaz. 24: 247-293.

Kavanaugh, D. H., 1986: A systematic review of Amphizoid beetles (Amphizoidae: Coleoptera) and their phylogenetic relationships to other Adephaga. – Proc. Calif. Acad. Sci. 44: 67-109.

Kocher, L., 1958: Catalogue commenté des Coléoptères du Maroc 2; Hydrocanthares, Palpicornes, Bracelytres. – Trav. Inst. scient. chérif., Sér. Zool. 14: 1-244.

Larsén, O., 1966: Locomotor organs of Gyrinidae and other Coleoptera. – Opusc. ent., Suppl. 3: 1-242.

Latreille, P. A., 1806: Genera Crustaceorum et Insectorum, etc., 1. – Paris.

– 1807: Genera Crustaceorum et Insectorum, etc. 2. – Paris.

– 1810: Considérations générales sur l'ordre naturel des Crustacés, Arachnides et Insectes. – Paris.

Lawrence, J. F., & A. F. Newton, Jr., 1982: Evolution and classification of beetles. – Ann. Rev. Ecol. Syst. 13: 261-290.

LeConte, J. L., 1850: General remarks upon the Coleoptera of Lake Superior, *in* Agassiz: Lake Superior. – Boston.

Lesne, P., 1902: Contribution a l'étude des premiers états des Gyrinides. – Bull. Soc. ent. Fr. 7: 85.

Lindberg, Harald, 1937: Finlands *Haliplus*-arter och deras utbredning inom Fennoscandia orientalis. – Acta Soc. Fauna Flora fenn. 60: 478-501.

Lindberg, Håkan, 1930: Zwei neue Fundorte für *Brychius elevatus*. – Notul. ent. 10: 123-126.

Lindroth, C. H. (ed.), 1960a: Catalogus Coleopterorum Fennoscandiae et Daniae. – Lund.

– 1960b: The larvae of *Trachypachus* Mtsch., *Gehringia* Darl., and *Opisthius* Kby. (Col. Carabidae). – Opusc. ent. 25: 30-42.

– 1985-86: The Carabidae (Coleoptera) of Fennoscandia and Denmark. – Fauna ent. scand. 15. Leiden/Copenhagen.

Linnaeus, C., 1758: Systema Naturae, ed. 10. – Holmiae.

– 1767: Systema Naturae, ed. 12. – Holmiae.

Lundberg, S., & A. N. Nilsson, 1978: Beetles (Ins.: Coleoptera) from the mouth-part of the river Ängerån. – Fauna Norrlandica 4: 1-8.

Meinert, F., 1901: Vandkalvelarverne (Larvae Dytiscidarum). – K. danske Vidensk. Selsk. Skr. (6), Mat. Nat. 9: 341-440.

Munster, T., 1927: Tillæg og bemærkninger til Norges koleopterfauna. – Norsk ent. Tidsskr. 2: 158-200.

Müller, O. F., 1776: Zoologiae Danicae Prodromus. – Havniae.

Neumayr, M., 1875: Über Kreideammonitiden. – Sber. Akad. Wiss. Wien, math. nat. Kl., 71: 639-693.

Motschulsky, V. de, 1853: Hydrocanthares de la Russie. – Helsingfors.

Ochs, G., 1930: Gyrinoidea. – Catalogue of Indian Insects 19. Calcutta.

– 1953: Ergebnisse der Österreichischen Iran-Expedition 1949-50. Gyrinidae (Coleoptera). – Sber. öst. Akad. Wiss., math. nat. Kl., Abt. 1, 162: 217-225.

- 1967: Zur Kenntnis der europäischen *Gyrinus*-Arten. - Ent. Bl. Biol. Syst. Käfer 63: 174-186.
Olivier, A. G., 1795: Entomologie ou Histoire Naturelle des Insectes, etc., Coléoptères 3 (41). - Paris.
Parry, J., 1982: The New Broom. - Balfour-Browne Club Newsletter 23: 1-2.
Paykull, G. von, 1798: Fauna Suecica, Insecta 1. - Uppsaliae.
Régimbart, M., 1883: Essai monographique de la famille des Gyrinidae 2. - Annls Soc. ent. Fr., Sér. 6, 3: 121-190.
Rousseau, E., 1920: Notes limnologiques 3; la larve présumée de *Brychius elevatus* Panz. - Bull. Soc. ent. Belg. 1: 165-167.
Ruhnau, S., 1985: Zur Morphologie und Biologie der praeimaginalen Stadien des Wasserkäfers *Noterus crassicornis* (Müller, 1776) (Coleoptera, Hydradephaga, Noteridae). - Diplom-Arbeit, Tübingen Universität.
Rutanen, I., 1976: A species of the genus *Gyrinus* L. (Col., Gyrinidae) new to Europe. - Annls Ent. Fenn. 42: 103-104.
Rydgård, M., *et al.*, 1985: Insektfaunan i en extremt sur sjö i Norrbottens kustland. - Ent. Tidsskr. 106: 133-138.
Satô, M., 1963: A new subgenus of the genus *Peltodytes* (Col., Haliplidae). - Ent. Rev. Japan 16: 21.
Saxod, R., 1964: L'oeuf, l'eclosion, la cuticule embryonnaire et la larve néonate de *Gyrinus substriatus* Steph. - Trav. Lab. hydrobiol. Grenoble 56: 17-28.
- 1965a: Cycle biologique de quatre espèces françaises de Gyrinidae. - Bull. Soc. zool. Fr. 90: 155-163.
- 1965b: Larves et nymphes de quatre espèces françaises de Gyrinidae. - Bull. Soc. zool. Fr. 90: 163-174.
Schiødte, J. C., 1864: De Metamorphosi Eleutheratorum observationes: Bidrag til Insekternes Udviklingshistorie. - Naturh. Tidsskr., 3. række, 3: 131-224.
- 1872: De Metamorphosi Eleutheratorum observationes: Bidrag til Insekternes Udviklingshistorie. - Naturh. Tidsskr., 3. række, 8: 165-226.
Seeger, W., 1971a: Morphologie, Bionomie und Ethologie von Hapliden, unter besonderer Berücksichtigung functionmorphologischer Gesichtpunkte (Haliplidae; Coleoptera). - Arch. Hydrobiol. 68: 400-435.
- 1971b: Autökologische Laboruntersuchungen an Hapliden mit zoogeographischen Anmerkungen (Haliplidae; Coleoptera). - Arch. Hydrobiol. 68: 528-547.
- 1971c: Die Biotopwahl bei Hapliden, zugleich ein Beitrag zum Problem der syntopischen (sympatrischen s. str.) Arten (Haliplidae; Coleoptera). - Arch. Hydrobiol. 69: 175-199.
Silfverberg, H., 1978: The coleopteran genera of Müller 1764. - Notul. ent. 58: 117-119.
- (ed.), 1979: Enumeratio Coleopterorum Fennoscandiae et Daniae. - Helsinki.
Svensson, B., 1969: The distribution and habitat specialization of the genus *Gyrinus* (Col. Gyrinidae) in Blekinge, southern Sweden. - Opusc. ent. 34: 221-242.
- 1982: *Gyrinus pullatus*, en för Sverige ny virvelbagge. - Ent. Tidsskr. 103: 55-60.
Tkačenko, M. I., 1929: Itinéraires des détachements zoologique de l'Expédition Iakoute de l'Académie des Sciences de l'URSS en 1926 et 1927. - Ezheg. zool. Muz. 30: 642-643. (In Russian).
Thompson, R. G., 1979: Larvae of North American Carabidae with a key to the tribes; *in* Ball *et al.* (eds.): Carabid Beetles: Their Evolution, Natural History, and Classification. - The Hague/Boston/London.
Vondel, B. J. van (in press): Description of the second and third instar larvae of *Haliplus laminatus* (Schaller) with notes on the subgeneric status (Coleoptera; Haliplidae).
Ward, R. D., 1979: Metathoracic wing structures as phylogenetic indicators in the Adephaga

(Coleoptera); *in* Ball *et al.* (eds.): Carabid Beetles: Their Evolution, Natural History, and Classification. - The Hague/Boston/London.
West, A., 1942: Fortegnelse over Danmarks Biller, etc. - København.
Zaitsev, F., 1915: Les coléoptères aquatiques de la collection Motschulsky, 1. Haliplidae, Dytiscidae, Gyrinidae. - Ezheg. zoól. Muz. 20: 239-295. (In Russian).
- 1953: Families Amphizoidae, Hygrobiidae, Haliplidae, Dytiscidae, Gyrinidae. - Fauna of the U.S.S.R., Coleoptera 4. Moskva/Leningrad. (In Russian; English translation Jerusalem 1972).
Zimmermann, A., 1920: Fam. Haliplidae *in* Junk & Schenkling (eds.): Coleopterorum Catalogus 4. - Berlin.
- 1924: Die Halipliden der Welt. - Ent. Bl. Biol. Syst. Käfer 20: 1-16, 65-80, 129-144, 193-213.
- 1925: Entomologische Ergebnisse der schwedischen Kamtschatka-Expedition 1920-1922, 9. Haliplidae und Dytiscidae. - Ark. Zool. 18B (5): 1-3.

Index

Synonyms and names of misidentified taxa are given in italics. The names in brackets refer to the genera in which the taxa are treated in this volume. The number in bold refers to the main treatment of the taxon.

Printed in the United States
By Bookmasters